国家林业和草原局职业教育"十三五"规划教材

木材利用基础（中英）

The Basis of Wood Utilization

卫佩行　龚　蒙　张　悦　主　编

马中青　关　鑫　副主编

王建和　赵广杰　主　审

中国林业出版社

内容简介

本教材内容涵盖木材宏观特征与识别、木材化学成分、木材多尺度细胞壁结构、木材水分、木材密度、木材力学性能等关系到木材利用的基本知识，融入思政，简明扼要，校企合作，突出应用，力求实现德技并修、学以致用之育人目的。

中英对照是本教材最显著的特点，也赋予了本教材在专业双语课程教学和国际交流中的优势。本书可作为高职院校的木业产品设计与制造、家具设计与制造等专业及应用型本科院校的相关专业的教材，也可作为企业工程技术人员的参考书。

图书在版编目（CIP）数据

木材利用基础 / 卫佩行，龚蒙，张悦主编 . — 北京：中国林业出版社，2021.6
（2024.9 重印）

国家林业和草原局职业教育"十三五"规划教材

ISBN 978-7-5219-1209-8

Ⅰ . ①木… Ⅱ . ①卫… ②龚… ③张… Ⅲ . ①木材－应用－高等职业
教育－教材 Ⅳ . ① S781

中国版本图书馆 CIP 数据核字（2021）第 109523 号

中国林业出版社·教育分社

责任编辑：田夏青　高兴荣
电　　话：（010）83143569

出　　版：中国林业出版社（100009　北京市西城区德内大街刘海胡同7号）
网　　址：http://www.forestry.gov.cn/lycb.html
发　　行：中国林业出版社
印　　刷：北京中科印刷有限公司
版　　次：2021 年 6 月第 1 版
印　　次：2024 年 9 月第 2 次
开　　本：787mm×1092mm　1/16
印　　张：15.75
字　　数：350千字
定　　价：65.00 元

本书可按需印刷，如有需要请联系我社。

《木材利用基础（中英）》
编写人员

主编

卫佩行　江苏农林职业技术学院

龚　蒙　加拿大纽布伦斯威克大学（University of New Brunswick）

张　悦　江苏农林职业技术学院

副主编

马中青　浙江农林大学

关　鑫　福建农林大学

参编人员

刘秀娟　江苏农林职业技术学院

张英杰　杨凌职业技术学院

翟　艳　山西林业职业技术学院

罗德宇　温州职业技术学院

陈　年　江西环境工程职业学院

孙丙虎　黑龙江林业职业技术学院

主审

王建和　宁波中加低碳新技术研究院有限公司

赵广杰　北京林业大学

前　言

职业教育作为一种类型教育，与普通教育具有同等重要的地位。为适应职业教育改革和发展之需，产教融合、校企合作人才培养模式下的教师、教材、教法"三教"改革应运而生。专业教材如何反映产业技术发展？在实际教学过程中，如何使用教材并采用合适的教法，实现既定的教学目的，这是教材编写者必须思考的问题之一。

据了解，国内尚无专门针对高等职业教育林业工程类专业的木材学相关教材。现有的普通高等教育《木材学》教材理论性强，学术性强，以培养研究型人才为目标；而高等职业教育以培养高素质技术技能型人才为目标，因而直接选用普通高等教育教材无法适应高职人才培养需要；其次，国内尚无木材利用方面的双语教材。当前，无论普通高等院校还是高等职业院校，为满足高等教育快速国际化和服务"一带一路"合作倡议，双语教学已经提到了一个新的高度。本书将是国内高等职业教育领域的首部木材学方面的双语教材，具有较好的示范效应。

为了适应应用型本科、职业本科及高等职业院校相关专业人才培养之需，本教材的编写特点为：

1. 简明扼要。教材内容密切结合生产实际，力求体现"够用、实用"原则。考虑到应用型本科生或高等职业院校学生的职业发展，本教材突出"是什么"的描述，避免过多"为什么"的介绍；另外，晦涩且在实践中应用极少的理论知识"点到为止"，不做过多阐述。

2. 融入思政。巧妙地将生态文明理念及节约资源、适材适用、大材优用、小材大用等木材利用理念融入教材内容。在传授专业知识的同时，更注重塑造学生的价值理念和精神追求。

3. 图文并茂。根据课程内容，适当提高插图和表格的比例，力求反映生产实际，使教材具有生动性和直观性。

4. 通俗易懂。教材编写贴近生产一线。对专业术语的解释准确规范，同时引入生产实例或通俗解释，以求理论与实践能够完美对接。

5. 分类选用。中文和英文部分可自成体系，根据学情选用；各个章节相对独立，可根据不同专业灵活选用。

6. 校企合作。在教材的编写过程中，广泛吸纳企业技术人员意见，力求实现既能用于在校学生知识传授又能用于在企员工技术培训之目的。

本教材由江苏农林职业技术学院卫佩行、加拿大纽布伦斯威克大学（University of New Brunswick，UNB）龚蒙、江苏农林职业技术学院张悦担任主编，由浙江农林大学马中青和福建农林大学关鑫担任副主编，由宁波中加低碳新技术研究院有限公司董事长、国际木材科学院院士、国家特聘专家王建和教授和北京林业大学赵广杰教授担任本教材主审。具体编写分工：第1章，张悦；第2章，卫佩行；第3章，关鑫、卫佩行、刘秀娟；第4章，罗德宇、陈年、卫佩行；第5章，卫佩行、张英杰、翟艳；第6章，卫佩行、张悦；第7章，龚蒙、卫佩行；第8章，马中青、孙丙虎。全书由卫佩行、龚蒙统稿。教材中部分插图由张悦和马琳绘制或修改。河南师范大学郭振胜对本教材有关内容提出了宝贵的修改意见。教材中文部分由咸阳师范学院中文系张乐妮通篇检查。

本教材英文部分由国际木文化学会专家润色提高，在此一并致谢。同时感谢俄罗斯MLT公司中国区负责人刘金、德国威力公司中国区销售经理朱祥光、圣象地板（句容）有限公司研发总监魏强、湖北福汉木业（集团）发展有限责任公司副总经理李文定博士提供的技术资料和修改建议。

本教材参考国内外木材科学、木材工程、木结构、林学、土木工程材料相关教材和参考文献，出版过程中得到了江苏农林职业技术学院刘玉华教授等领导和中国林业出版社领导与编辑的大力支持。在此，向本教材编写、编辑、出版和发行等做出贡献的同仁和出版工作者表示衷心的感谢。

本教材不仅可作为木业产品设计与制造、家具设计与制造、建筑工程技术、园林工程技术、林业技术等专业的教材，还可以作为环境艺术设计等专业的教材或参考书。对于木材加工、家具制造、木结构建筑等从业人员，本教材也是很好的参考书。

本教材编写既要体现高等职业教育特色和专业特色，又要保证英文表达准确和专业；既要反映行业发展新动态、新技术、新产品、新工艺，又要兼顾学生知识结构，因而本教材的编写经历了艰苦的探索过程。由于编者自身水平和知识结构都有局限性，疏漏和不足在所难免。因此，恳请本教材的使用者能够提出严厉批评和宝贵意见，以便修正、完善！另外，本教材同步开发了课程资源，需要者可联系编者。编者联系方式：wayne0448123@163.com。

卫佩行
2020 年 8 月

Preface

As a type of education, vocational education plays an equally important role with general education. In order to meet the needs of the reform and development of vocational education, the reform of teacher training, textbooks, and teaching methods came into being under the talent training mode of integration of industry and education and school–industry collaboration. In this case, how the textbooks reflect the development of industrial technology and how to use textbooks and select appropriate teaching methods during the teaching to achieve the established teaching objectives has become the problems that textbook editors must consider.

As far as we know, there are few textbooks about wood science for forestry engineering majors in tertiary vocational education in China. The existing textbooks of wood science for undergraduate are highly theoretical and academic, since undergraduate education aims at cultivating research–oriented talents, which are unsuitable for tertiary vocational colleges with an aim to cultivate high–quality technical or application–oriented talents. In addition, there are few bilingual textbooks on wood utilization in China. At present, with the development of the Belt and Road initiative and the internationalization of higher education, bilingual instruction has become a new trend in colleges and universities. After the publication of this textbook, it will be the first bilingual textbook on wood science in the field of tertiary vocational education in China, which has a good demonstration effect.

In order to meet the needs of talent training for application–oriented undergraduate and vocational undergraduate in colleges, the compiling principles of this textbook are as follows:

1. Clarity. This textbook, with sufficient and practical content, are closely related to the actual production process. Considering the career development of students, this textbook highlights the description of "what is it" while avoiding too much introduction of "why" . In addition, it does not elaborate too much on the obscure theoretical knowledge which is rarely applied in practice.

2. Integration of ecological concepts. The concepts of ecological civilization and wood utilization such as resource conservation are added into this textbook, which means that shaping students' values are also paid attention while imparting professional knowledge.

3. Illustrated. This textbook appropriately added the proportion of figures and tables to reflect the actual production based on the content, so that the textbook is vivid and intuitive.

4. Easily understood. This textbook is closely related to actual production process. For example, while explaining professional terms, production examples, and popular interpretation are introduced to the textbook to help students understand them better.

5. Optional. The Chinese and English parts of this textbook are self—contained which means that they can be selected according to the learning situation of certain students. In addition, each chapter is also relatively independent and can be chosen flexibly according to different disciplines.

6. School—enterprise collaboration. During the writing, we widely absorb the opinions of technicians from enterprises, and strive to achieve the purpose of both imparting knowledge to students in school while doing the technical training for employees in enterprises.

There are three chief authors, including Dr. Peixing Wei from Jiangsu Vocational College of A&F, China, Dr. Meng Gong from the University of New Brunswick (UNB), Canada, and Ms. Yue Zhang from Jiangsu Vocational College of A&F. Two deputy authors—in—chief, Dr. Zhongqing Ma and Dr. Xin Guan, are respectively from Zhejiang A&F University and Fujian A&F University. Prof. Brad Jianhe Wang, chairman of Ningbo Sino—Canada low carbon New Technology Research Institute Co., Ltd, fellow of International Academy of Wood science, and National distinguished expert of China, and Prof. Guangjie Zhao, professor of Beijing Forestry University, serve as the chief reviewers of this textbook. This textbook was summarized, revised and finalized by Peixing Wei and Gong Meng. Some figures in the textbook were drawn or modified by Yue Zhang and Lin Ma. Dr. Zhensheng Guo from Henan Normal University put forward valuable suggestions on revising the relevant contents of this textbook. The Chinese part of this textbook was examined by Leni Zhang from Xianyang Normal University. The English part of this textbook was polished and improved by experts of International Wood Culture Society. Thank these experts for their contributions! I also would like to thank Jin Liu, the manager of MLT, Zhu Xiangguang, the sales manager of Weinig, Qiang Wei, the R & D director of Power Dekor flooring (Jurong) Co., Ltd., and Wending Li, the deputy general manager of Fuhan wood industry (Group) Development Co., Ltd., for their technical documents and suggestions! In addition, I must apologize that some photos and materials were downloaded from the Internet which cannot be verified or difficult to trace the sources.

This textbook referred to relevant textbooks and references of wood science, timber engineering, wood construction, forestry and civil engineering at home and abroad. Yuhua Liu and other staff from Jiangsu Vocational College of A&F and the leaders and editors of China Forestry Publishing House were extremely helpful and supportive in getting this book published. I would like to express my gratitude to the experts, colleagues, leaders and publishers who have made contributions to the writing, editing, and publishing of this textbook!

This textbook can not only be used as a textbook for forestry and civil engineering majors such as wood product design and manufacturing, furniture design and manufacturing, construction engineering technology, garden engineering technology, and forestry technology, but also as a reference for majors such as international trade, interior design, and environmental art design. This textbook is also a good reference for practitioners engaged in wood processing, furniture manufacturing, and wood construction.

The textbook should not only reflect the characteristics of tertiary vocational education and wood industry, but also ensure the accuracy and idiomatic expression of English. It needs to reflect the new trends, new technologies, new products and new processes of industry development while at the same time building a knowledge structure for students. Therefore, the compilation of this textbook has experienced a difficult exploration process. The authors believe that the development of science and technology is changing with each passing day, and the content of textbooks inevitably lag behind. Furthermore, due to the limitations of the authors' level and ability, omissions and deficiencies are inevitable. Therefore, the users of this textbook are kindly appreciated for severe criticism and valuable opinions! In addition, the authors of this textbook had developed curriculum resources simultaneously, and those who need it can contact me via: wayne0448123@163.com.

Peixing Wei

2020.8

目 录

前　言

第1章　森林与木材 1

1.1　森林与人类 2

1.2　木材与生活 7

1.3　木材的特性 16

第2章　木材来源 19

2.1　植物分类与木材来源 19

2.2　树木生长与木材形成 24

2.3　木材的命名 29

第3章　木材构造 31

3.1　木材的主要宏观构造 32

3.2　木材的主要显微构造 37

3.3　木材识别方法 41

3.4　主要商品材介绍 43

第4章　木材细胞壁 51

4.1　木材细胞壁主要成分 51

4.2　木材细胞壁结构 56

4.3　木材化学利用举例 58

第5章　木材物理性质 60

5.1　木材水分相关性质 60

5.2　木材密度和孔隙度 68

5.3　木材的其他物理性质 71

第6章　木材力学性质 ·············· 75

6.1　木材力学基础 ················· 75

6.2　木材弹性常数 ················· 78

6.3　木材主要力学性能指标 ············· 79

6.4　木材力学性质的影响因素 ············ 85

第7章　木质材料 ··············· 88

7.1　木材锯解 ·················· 88

7.2　锯材干燥 ·················· 90

7.3　木质人造板加工 ················ 92

第8章　木材缺陷 ·············· 106

8.1　木材天然缺陷 ················ 106

8.2　生物危害缺陷 ················ 111

8.3　木材加工缺陷 ················ 113

参考文献 ·················· 235

CONTENTS

Preface

Chapter 1   Forest and Wood ·· 117

1.1   Forest and Human Survival ·· 117

1.2   Wood and Human Living ·· 124

1.3   Characteristics of Wood·· 131

Chapter 2   Sources of Wood ·· 134

2.1   Plant Taxonomy and Wood Products ·· 134

2.2   Tree Growth and Wood Formation·· 136

2.3   Nomenclature of Wood ·· 141

Chapter 3   Structure of Wood ·· 143

3.1   Macroscopic Characteristics of Wood ·· 143

3.2   Microscopic Characteristics of Wood·· 149

3.3   Wood Identification ·· 155

3.4   Introduction of Main Commercial Woods ·· 157

Chapter 4   Structure of Wood Cell Wall ·· 166

4.1   Chemical Components of Wood Cell Wall ·· 166

4.2   Cell Wall Structure of Wood ·· 171

4.3   Examples of Wood Chemical Utilization ·· 174

Chapter 5   Physical Properties of Wood ·· 175

5.1   Wood Water and Density ·· 175

5.2   Other Physical Properties of Wood ·· 187

Chapter 6　Mechanical Properties of Wood ·················· 192

6.1　Basic Concepts of Wood Mechanics ···················· 192

6.2　Wood Elastic Constants ·································· 194

6.3　Main Mechanical Properties of Wood ················· 195

6.4　Influencing Factors of Wood Mechanical Properties ·········· 202

Chapter 7　Wood-based Materials ························· 205

7.1　Conversion of Timber ································· 205

7.2　Timber Seasoning ···································· 207

7.3　Wood-based Products ································· 210

Chapter 8　Wood Defects ······························· 226

8.1　Wood Natural Defects ································ 226

8.2　Wood Biohazard Defects ···························· 229

8.3　Wood Artificial Defects ····························· 232

References ··· 235

第❶章
森林与木材

⟫ **学习目标**

（1）了解森林资源现状；

（2）了解森林的生态功能；

（3）了解木材在生活中的主要应用；

（4）了解木材的特性。

⟫ **本章描述**

　　森林不仅具有重要的生态功能，还为人类提供重要的木材资源。随着生态文明建设的持续推进，木材资源的重要性将进一步凸显。为充分利用珍贵的木材资源，有必要深入了解森林保障人类生存的重要性，以及木材的主要特性和主要应用领域。

　　森林（forest）作为地球上结构最复杂、功能最多和最稳定的陆地生态系统，被誉为大自然的"总调节器"和"地球之肺"，在维持生态平衡、促进人与自然和谐共生、护佑人类生存与发展中具有决定性和不可替代的作用。

　　森林除生态功能外，还提供重要的**木材**（wood）[①]资源。在水泥（cement）、钢铁（steel）、木材和塑料（plastic）四大建筑材料中，木材是唯一可再生的生态材料。随着现代工业发展和科技进步，木材用途将越来越广泛。

　　然而，木材由于常见易得，往往得不到人们的珍惜。长期的野蛮开发和不合理利用，使得森林资源急剧减少，并带来了前所未有的生态危机。尽管人们重新认识到人类的生存发展与森林生态系统的密切关系，但是人们对木材的珍惜程度以及科学合理使用的重要性依然认识不足。保护森林资源，合理使用木材，首先要具有敬畏自然、节约木材的正确心态。

---

① 当"锯材"使用时，既可以使用"timber"也可以使用"lumber"。木材的宽厚比小于 2 时用"timber"，宽厚比大于 2 时用"lumber"；当"木材"使用时，某种程度上"lumber"和"wood"可以互换，但两者在外延上略有不同，"lumber"是指具有各种天然缺陷的试验用木材，而"wood"通常是指无缺陷、纹理通直的无疵木材。"dimension lumber"相对于"timber""lumber"和"wood"三个词，其含义较为具体，是指应用于轻型木结构的对尺寸和性能有一定要求的锯材。

1.1　森林与人类

1.1.1　森林资源现状

森林的种类多样。根据起源，森林可分为**天然林**（natural forest）和**人工林**（planted forest）。根据用途，森林可分为 5 个森林种类，简称"林种"。它们分别是：用材林、防护林、经济林、薪炭林和特种用途林（图 1.1）。

我国森林的界定标准为：**郁闭度**（canopy density）①在 0.2 以上；天然林②面积要达到 1.5 亩③以上，而人工林④、**经济林**⑤则要达到 1 亩以上。

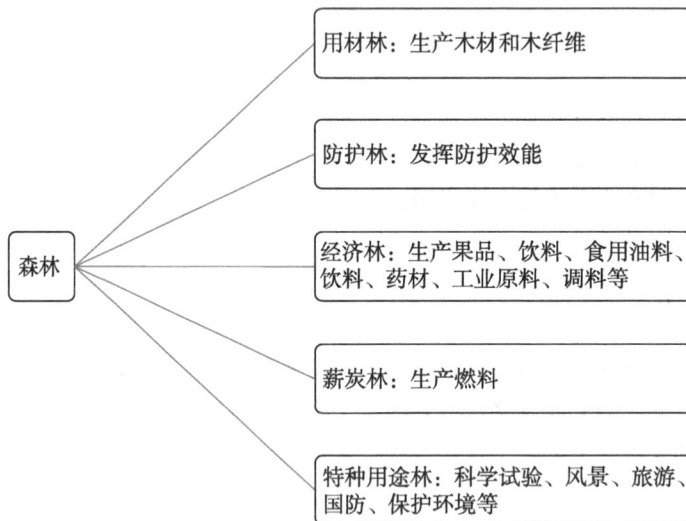

图中文字：

用材林：生产木材和木纤维

防护林：发挥防护效能

经济林：生产果品、饮料、食用油料、饮料、药材、工业原料、调料等

薪炭林：生产燃料

特种用途林：科学试验、风景、旅游、国防、保护环境等

森林

图 1.1　基于用途的森林分类

全球森林总面积约 40.6 亿 hm²，覆盖率约 30.8%，其中针叶林⑥占 1/3。然而，全球森林的分布极不均衡，森林资源最丰富的 5 个国家（俄罗斯、巴西、加拿大、美国和中国）占有全球森林总面积的一半以上。在全球范围内，天然林面积占世界森林面积的 93%。森林的 45% 位于热带地区（最大部分），其次是亚寒带、温带和亚热带地区。

根据第九次中国森林资源清查（2014—2018 年），我国现有森林面积为 2.2 亿 hm²，

① 郁闭度指森林中乔木树冠在阳光直射下在地面的总投影面积（冠幅）与此林地（林分）总面积之比，它反映林分的密度。联合国粮农组织规定，0.70 以上（含 0.70）的郁闭林为密林，0.20~0.69 为中度郁闭，0.20 以下（不含 0.20）为疏林。

② 天然林指天然起源的森林，包括自然形成与人工促进天然更新或者萌生所形成的森林。

③ 1 亩 ≈ 667m²。

④ 人工林指通过人工措施形成的森林。

⑤ 经济林亦称"特用林"，是指以生产果品、食用油料、工业原料和药材为主要目的的林木。

⑥ 针叶林是以针叶树为建群种所组成的各类森林的总称。

森林蓄积量为 175.6 亿 m³，森林覆盖率为 22.96%。我国是世界上森林树种最多的国家，特别是珍贵稀有树种。据我国植物学家统计，我国有种子植物 2 万余种，其中属于森林树种的有 8000 余种。

属于针叶树的松、杉树种，是北半球的主要树种。全球约有 30 属，而我国就有 20 属，近 200 种，其中 8 个属是我国特有，即水杉、银杉、金钱松、水松、台湾杉、油杉、福建柏和杉木。

阔叶树种则更加丰富，有 200 属之多，其中许多是我国特有树种，如珙桐、杜仲、喜树、香果树和樱椒树等。

与世界其他各国和地区森林资源相比，我国森林资源主要的不足体现在：

（1）森林资源少，覆盖率低

我国森林覆盖率大约只有全球平均水平的 2/3，人均森林面积不足世界平均水平的 1/4，人均森林蓄积量只有世界平均水平的 1/7。

（2）森林资源分布不平衡

全国绝大部分森林资源集中分布于东北、西南等偏远山区和台湾山地及东南丘陵，而广大的西北地区森林资源贫乏。

◎ 东北林区是我国最大的天然林区，森林资源主要集中在大兴安岭、小兴安岭和长白山地，主要的树种有红松、兴安落叶松、黄花落叶松等针叶树，也有白桦、水曲柳等阔叶树；

◎ 西南林区是我国第二大天然林区，主要分布在横断山脉，主要树种有云杉、冷杉、高山栎、云南松等，还有珍贵的柚木、紫檀、樟木等；

◎ 东南林区是我国主要的经济和人工林区。这里气候温暖，降水充沛，植物生长条件良好，树木种类很多，以杉木和马尾松为主，还有我国特有的竹林。

（3）用材林多，防护林少

各林种比例不够合理，难以发挥森林资源的综合效益。

（4）森林资源质量不高

林业用地有林地① 比例低，单位面积蓄积量和森林生长率也低。我国现有森林中，原始林少，天然残次林多，林相② 差。森林可采资源少，木材供需矛盾突出。

1.1.2 森林的生态功能

（1）制氧固碳

每一棵树都是一个**氧气"发生器"**和**二氧化碳"吸收器"**。森林植物通过光合作用吸收二氧化碳，释放氧气，把大气中的二氧化碳以生物量的形式固定在植被和土壤中。森林每增加 1m³ 的蓄积量，平均多吸收 1.83t 二氧化碳，多释放 1.62t 氧气。

———————————

① 有林地（forest land）指树木郁闭度大于等于 0.20 的林地。
② 林相就是"森林的外形"。一是指林冠的层次，分为单层林和复层林；二是指森林的林木品质和健康状况。林木价值较高，生长旺盛，称为林相优良，反之则称为林相不良。

森林是陆地上最大的"**储碳库**"和最经济的"**吸碳器**"。全球森林面积虽然只占全球陆地总面积的 1/3，但每年吸收的二氧化碳却占生物固碳总量的 4/5。

森林所固定的碳，可以长期存在于树木中，更可以长久保存于各种木制品和林产品中。因此发展林业产业，加快森林资源培育，增强森林的碳汇功能，已成为全球的共识和期望（图 1.2）。

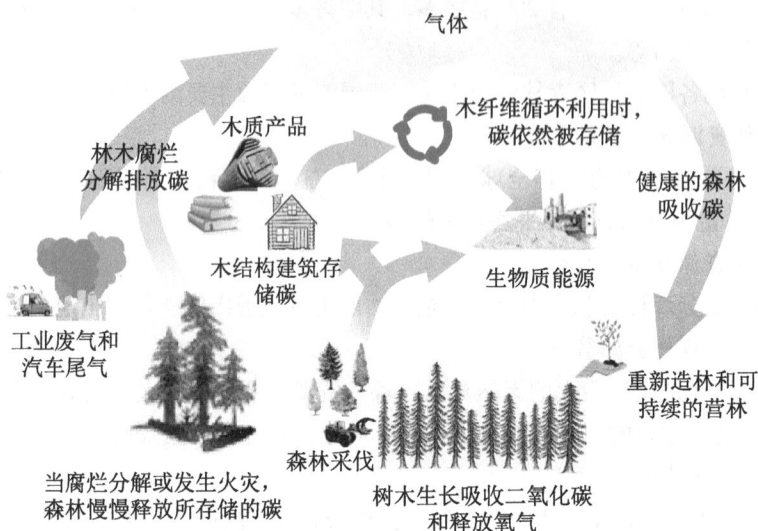

气体

木纤维循环利用时，
碳依然被存储

木质产品

林木腐烂
分解排放碳

健康的森林
吸收碳

木结构建筑存
储碳

生物质能源

工业废气和
汽车尾气

重新造林和可
持续的营林

当腐烂分解或发生火灾，
森林慢慢释放所存储的碳

森林采伐

树木生长吸收二氧化碳
和释放氧气

图 1.2　森林碳循环

（2）调温增湿

森林是天然的"空调"。在浓密的林冠[①]下，地表与大气之间形成一个"**绿色调温器**"，它不仅使林内环境有特殊的变化，而且对森林周围的温度也有很大的影响。与无林地[②]相比，林内冬暖夏凉、夜暖昼凉，温差较小。

（3）保持水土

森林具有复杂的复层结构和良好的地表覆盖。强大的植物根系和土壤渗透系统，能有效地截留降水的地表径流，防止和减少水土流失。研究表明，$1hm^2$ 森林可贮 $300\sim1000m^3$ 的水。营造 1 万 hm^2 的森林，相当于修建一座库容 300 万 ~1000 万 m^3 的水库。森林通过林冠、林地枯落物[③]截留降水、保持水分和土壤调节（图 1.3），实现对降水的再分配和净化。森林不仅能涵养水源，贮蓄水分，还能降低水的硬度，提高水的碱性，有效改善水质。

（4）改良土壤

树木根系的穿透能改善土壤的物理结构。林地的枯枝落叶、种子、芽、树皮等残落

---

① 树木上部着生的全部枝和叶称为树冠。在林分中，树冠层的总和称为林冠。

② 无林地（non-stocked land）指尚未绿化的林业用地。虽有零星散生木，但其郁闭度小于 0.1 的林地，以及造林成活率严重不足的林地，均划作无林地。

③ 枯落物是植物地表器官枯死后的所有有机质的总称，地表枯落物蓄积量与径流等因素密切相关。

物、死地被物①和动物尸体，在风、水、阳光、微生物和各种动物的作用下，能分解成肥力很高的腐殖质②，增加土壤有机质和植物生长所需的氮、磷、钾等元素的含量，从而提高土壤肥力。同时，森林中有许多鸟类、兽类，其粪便对肥沃森林土壤也发挥着很大的作用。植物通过根系还可以吸收土壤中的镉、铅、铜、锌、汞等重金属，减轻土壤重金属的污染（图 1.4）。

图 1.3　森林涵养水源

图 1.4　森林改良土壤

（5）杀菌消毒

森林能分泌**杀菌素**③，如萜烯、乙醇、有机酸、醚、醛、酮等。这些物质能杀死细菌、真菌和原生动物，使森林中空气含菌量大大减少。据监测，$1hm^2$ 阔叶林 24h 内能产生植物杀菌素 2kg，而针叶林为 5kg 以上。

（6）净化空气

许多植物能在保持正常生理机能的状态下吸收大气中的污染物质，并在体内代谢、降解或富集，使大气污染得到一定程度的净化。如二氧化硫是大气主要污染物之一。而硫却是树木体内氨基酸的组成成分，也是树木生长所需要的营养元素之一。

（7）滞留尘烟

一方面由于森林中树木的枝叶茂密，可以阻挡气流和减低风速，使尘埃在大气中失去移动的动力而降落；另一方面，树木叶片有一个较强的蒸腾面，晴天要蒸腾大量水分，使树冠周围和森林表面保持较大湿度，使尘烟吸水增加重量从而较易降落吸附，雨天树木叶片的尘烟被雨水淋洗，空气经过森林反复洗涤过后，便会变得清洁；再一方面，树木的花、果、叶、枝等能分泌多种黏性汁液，同时表面粗糙多毛，会使空气中的尘烟被黏着、阻滞和过滤。

（8）隔离噪声

树木粗糙的树干、茂密的枝叶以及林下的枯枝落叶层有反射、吸收和阻隔噪声的作

---

① 死地被物指林地上的枯枝落叶层。包括枯倒的林木、下木、幼苗、幼树、层外植物、活地被物的凋落物和动物残骸以及排泄物等。它们是土壤腐殖质和肥力的来源。

② 腐殖质是指新鲜有机质经过微生物分解转化所形成的黑色胶体物质，一般占土壤有机质总量的 85% 以上。

③ 杀菌素指树木分泌的具有抑菌杀菌能力的一种挥发性物质。能直接杀死细菌、真菌、原生动物。

用。一条宽 40m 的乔木林带可降低噪声 10~15dB，最高达 30dB；宽 20m 的乔木林带可使噪声降低 8~10dB。城市中绿化街道比无绿化的能使噪声降低 3~10dB。公园内成片的林木能使噪声降低 26~34dB（图 1.5）。

（9）医疗保健

森林通过光合作用可产生大量氧气，提高空气中氧的含量，呼吸这些新鲜空气能清肺强身。另外，树木、枝叶尖端放电及绿色植物光合作用形成光电效应[1]，使空气电离而产生空气负（氧）离子[2]。森林里每 1cm^3 空气中含负离子高达 2 万个以上。空气负离子又称 "**空气维生素**"，它对各种细菌、病毒产生较强的抑制作用，还可通过呼吸道经肺部进入人体血液，促进血液循环，提高人体的免疫力，具有一定的医疗保健作用（图 1.6）。

图 1.5　森林隔离噪声

图 1.6　森林保健

（10）保护生物多样性

森林不仅分布区域广、自然地理环境类型繁多，而且是地球上结构最复杂和最稳定的陆地生态系统，食物链完整而复杂，不但为各种植物和微生物提供了生存的基础和营养来源，也为动物提供了栖居场所和丰富的食物（图 1.7）。

图 1.7　森林保护生物多样性

---

① 光电效应是在高于某特定频率的电磁波（该频率称为极限频率）照射下，某些物质内部的电子吸收能量后逸出而形成电流，即光生电。

② 空气负（氧）离子（negative air ion，NAI）是带负电荷的单个气体分子和轻离子团的总称。森林和湿地是产生空气负（氧）离子的重要场所。

科学家经测算发现，人类正常的衣食住行约需要 4 万种生物来维持。而每一种物种的绝迹，都预示着很多物种即将面临灭绝。科学家观察发现，一种生物的消失会引起相关联的 20 个物种的消失。因此，我们要呵护自然，保护森林，保护生物多样性。

1.1.3　森林开发利用新方向——森林康养

森林环境能在一定程度上减少肾上腺素的分泌，降低交感神经的兴奋性。经常处在优美、安静的绿色环境中，如在森林公园内游览，人的皮肤温度可降低 1~2℃，脉搏恢复率可提高 2~7 倍，每分钟脉搏次数能减少 4~8 次，呼吸慢而均匀，血流减缓，心脏负担减轻，听觉和思维活动的灵敏性增强近一倍。

在森林环境里，人们可以进行攀缘、徒步、观光、探秘、露营、疗养等多项活动。预计，在 21 世纪，以森林旅游为主要的生态旅游是旅游业中发展速度最快的部分，将以 30% 增长率发展，全球旅游者将有一半以上走进森林（图 1.8、图 1.9）。

图 1.8　森林徒步

图 1.9　森林疗养

1.2　木材与生活

我们的远古祖先从**钻木取火**，到**制木为器**，再到**以木为居**，逐步建立了人类文明。可以说，木材在人们的衣食住行等各方面均起着举足轻重的作用。即使在科技、经济发达的今天，木材仍旧在人们的生产、生活中扮演着不可替代的角色。

1.2.1　木文化

人类的发展历史也是木材的利用史。作为我国第一部诗歌总集，《诗经》也集中展现了远古时期我国缤纷独特的木文化，以及先民与树木之间休戚与共的亲密关系。据统计，《诗经》共 305 篇，其中有 79 篇涉及 61 类木本植物：大乔木与乔木 26 类、灌木 29 类、木质藤本 6 类。

在汉字中木是构字的重要元素之一（图 1.10）。"木"字本身就是树木的象形文字，**中间的一竖是树干，上面的一横是树枝，下面的一撇一捺是树根**。一木为树，两木为林，三木为森。与树有关的字均以木为偏旁，如柳、松、杉、樟等。据统计，《诗经》中涉及

图 1.10 "木"字的演变

"木"的字有 132 个。

总而言之，**木文化是人类和自然木在漫长的对话、交流的过程中形成的有形质的凝固态或无形质的非凝固态的"人化"木，是物质和精神的融合**。木材利用既显示了功能性，又体现了艺术性，同时反映了人类对木材的认知和理解水平。而木质文创产品是对当代木文化的最好诠释，包含木质工艺品和衍生文创产品两大类。

在木质工艺品中，木雕最具代表性。由于民俗、文化、工艺和资源的不同，我国形成了诸多具有浓郁地方特色、各有千秋的木雕流派。其中，最为著名的木雕流派为东阳木雕、乐清黄杨木雕（图 1.11）、广东潮州金漆木雕和福建龙眼木雕，这四大流派被称为**"中国四大木雕"**。

衍生文创产品（图 1.12~ 图 1.14）集实用性、观赏性、趣味性于一体，主要包括文化生活类、文博旅游类、动漫影视类等，木质材料的自然文化属性可以提升产品的内涵品质，木质材料的感官体验由此产生情感共鸣。

图 1.11 黄杨木雕凤凰香插

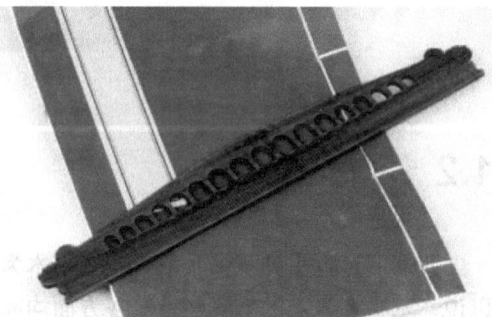

图 1.12 十七孔桥御尺（颐和园）

图 1.13 令牌木质书签（中国国家博物馆）

图 1.14 木质抽纸盒

1.2.2　木建筑

木建筑是中华民族五千年文明在居住方面最杰出的载体。从远古先人的凿巢穴居，夯土筑屋，到美轮美奂、精益求精的飞檐翘角、雕梁画栋，中国的古代建筑延续数千年之久，独树一帜。

榫卯（mortise and tenon）〔图 1.15（a）〕是古代中国建筑、家具及其他器械的主要结构形式，是在两个构件上采用凹凸部位相结合的一种连接方式。凸出部分叫**榫**（或叫榫头），凹进部分叫**卯**（或叫榫眼、榫槽），其特点是在构件上不使用钉子。利用榫卯加固构件，体现出中国古老的文化和智慧。

斗拱（bracket set）〔图 1.15（b）〕，是中国古代建筑特有的一种**榫卯**结构。在立柱和横梁交接处，从柱顶上加的一层层探出成弓形的承重结构叫**拱**，拱与拱之间垫的方形木块叫**斗**，合称**斗拱**。

（a）榫卯　　　　　　　　　　　　　　（b）斗拱

图 1.15　榫卯与斗拱

据考古发现，河姆渡遗址中的干阑式建筑（图 1.16）是我国最早发现的木结构，使用的就是榫卯结构。这种建筑以竹木为主要建筑材料，主要是两层建筑，下层放养动物和堆放杂物，上层住人。我国古代建筑种类丰富，包含寺观、宫殿、坛庙、佛塔、陵墓、园林建筑和民居等。五台山南禅寺（图 1.17）是世界留存至今最古老的木结构佛寺。北京故宫（图 1.18）是世界上现存规模最大、保存最为完整的木结构古建筑之一。北京天坛祈年殿（图 1.19）是木结构坛庙建筑的杰出代表。

图 1.16　河姆渡遗址干阑式建筑模型　　　　图 1.17　南禅寺大佛殿

图1.18 故宫太和殿

图1.19 天坛祈年殿

从辽代留下的山西应县佛宫寺释迦塔（应县木塔）（图1.20），距今（2021年）已有965年，塔高67.31m，底层直径30.27m，总重量约7400t。整个建筑由塔基、塔身、塔刹3个部分组成。塔基分为上下两层，下层为方形，上层为八角形。塔身平面亦为八角形，塔高9层，5个明层和4个暗层，外观为5层六檐。木塔通体全靠木构件榫卯咬合。上层柱脚插在下层柱头的枋上，并向内递收，形成一层比一层小的优美轮廓。全塔共使用54种不同形式的斗拱，种类之多，国内罕见，被世人称为"斗拱博物馆"。

位于江苏扬州的汉广陵王墓属于"**黄肠题凑**"式木椁墓（图1.21），规模宏大，结构

图1.20 山西应县木塔

严谨，是全国罕见的大型汉代墓葬之一，距今已有两千多年的历史。"黄肠"指的是所用木料为柏木心材，以木色淡黄而得名。"题凑"中的"题"指的是柏木的头（靠近树冠的一头）；"凑"，以头向内。合起来解释，"黄肠题凑"是指用心材柏木，按向心方式堆垒而成的厚木墙。明长陵祾恩殿（图1.22）是中国为数不多的大型楠木殿宇，汇集了古代建筑的特点，展示了古代劳动人民在建筑艺术上的无穷智慧。清东陵定东陵隆恩殿（图1.23）的梁枋架木、门窗隔扇，全部采用名贵的黄花梨木，堪称"木绝"。

图1.21 汉广陵王墓与"黄肠题凑"

图 1.22　明长陵祾恩殿

图 1.23　清东陵定东陵隆恩殿

由于大径级的天然林木材减少，现代木结构建筑（图 1.24）主要用材为人工林小径材加工而成的工程木材。连接采用的是各类金属连接件，其结构稳定性、耐用性优于传统木结构，也拓展了木结构建筑的设计跨度、高度及广度。预制化、装配化程度极高的施工方法，能够更好地保证施工质量并提高施工速度。

（a）轻型木结构

（b）木刻楞结构

（c）梁柱式结构

（d）高层木结构

图 1.24　现代木结构建筑

1.2.3　木家具

建筑在表、家具在内。中国传统家具正是工匠们用木材创制的绚丽诗篇。尤其明清家具，是中国传统文化的瑰宝，代表了最具中国传统风格的古典家具。

图 1.25（a）为经典明式圈椅。圈椅造型为上圆下方，外圆内方。背板做成"S"或"C"形曲线，是根据人体脊椎骨的曲线制成的，为明式家具科学性的一个典型例证。

图 1.25（b）为新中式家具。新中式家具是在传统美学规范之下，运用现代的材质及工艺，演绎传统中国文化的精髓，使家具不仅拥有典雅、端庄的中国气息，并具有明显的现代特征。

图 1.25（c）为路易十五式扶手椅。路易十五式扶手椅是典型的洛可可风格，整体由流畅的曲线构成，敞开式扶手，且向里缩进一段，靠背和坐面包覆色彩淡雅秀丽的织锦。脚端常雕刻成猫脚或山羊脚，纤细柔美。

图 1.25（d）为三脚贝壳椅。三角贝壳椅因其椅面曲线优美，好似温暖的微笑，别名"微笑椅"。它的微笑座面呈现出独特的三维曲面效果，如羽翼般轻盈流畅，悬浮灵动。

（a）经典明式家具（圈椅）

| （b）新中式家具
（中国椅） | （c）西方古典家具
（路易十五式扶手椅） | （d）西方现代家具
（三角贝壳椅） |

图 1.25　古今中外典型家具

1.2.4　木材装饰

在建筑装饰工程中，木材作为天然生态的饰面材料得到广泛应用，包括水平方向的地面及顶棚、垂直方向的墙面和柱面、门窗、木线条等。

木材在现代建筑地面装修上应用，最显著的优势就是导热系数小，可以缓和外部气温变化所引起的室内温度变化。另一个优势是木材弹性和硬度均适中，是运动场馆、演出舞台等对地面有特殊力学要求场所的最佳选择（图 1.26）。

（a）幼儿园木地板　　　　　　　　　　（b）篮球馆木地板

图 1.26　地面装饰材料——地板

木材重量轻但强度大，应用在顶棚装修时能够很好地满足构造的力学要求，还不会给建筑基体带来太多的附加荷载。因此，在室内装修中用木材作为悬吊式顶棚的龙骨很常见，并且木材的光学、声学性能独特，作为顶棚材料能够为室内环境的塑造起到独特的作用（图 1.27）。

图 1.27　木材应用于室内顶棚和墙面

木材在墙面中应用的优势主要是其对室内环境的调节能力：由于木材具有湿度调节能力，应用于墙面会减少墙面的结露；导热系数小能够提高室内环境的保温性；天然多孔性能够使其表面产生光线的漫反射，使令人眩晕的光线变得柔和，改善室内的光环境；利用木材对声音的反射和吸收，改善室内的声环境。另外，木材因其硬度适中，应用于室内墙面还具有缓冲特性，在外力的撞击下能够很好地保护墙体和人（图 1.28）。

木材在门窗中应用可以防止冷热桥的产生，为室内环境提供隔热保温的作用。木质线脚、踢脚板均具有掩盖缝隙的作用。由于其质轻高强、加工性好、握钉性好，实木还能够在现代建筑室内装修中作为具有辅助功能的构件。

（a）中国哈尔滨大剧院室内装饰

（b）挪威野生驯鹿瞭望亭室内装饰

图 1.28　木材装饰室内空间

1.2.5　木材药用

　　许多木材拥有极高的药用价值。这些木材的药用价值在几千年前的中医药中就有所应用，并沿用至今（图 1.29）。如从降香黄檀的心材中提取的"降香"具有抗肿瘤、保护心血管、抗炎消炎、抗自由基氧化、镇静安神等作用。降香经过蒸馏后提取的降香油可做镇痛剂使用，常在中医药中将降香和其他药搭配使用。因其具有降血压的功效，降香黄檀被誉为"降压木"。

图 1.29　中药铺

1.2.6 木材能源

采用现代科学技术，可以将木材加工成高效方便的能源。常用以下 3 种木材能源化加工方法：木材液化（liquefaction）生产燃料油、木材水解（hydrolysis）生产燃料乙醇，木材热解（pyrolysis）气化（gasification）发电。

木材是由碳、氢、氧构成的高聚物，而碳、氢、氧是燃料物质的重要组成元素。通过物理、化学或生物的方法，可以将木材液化成燃料油。常见的木材液化方法包括高温高压液化、常压催化剂液化、超临界法液化、微生物助解液化，将木材转化成生物柴油。

木材主要由"三大素"组成，即纤维素（cellulose）、半纤维素（hemicellulose）和木质素（lignin）。其中纤维素和半纤维素多为糖类（sugars）物质，约占木材总量的 75%，糖类物质经过发酵生成乙醇（ethanol）。用木材生成乙醇的关键是木材中高聚糖的水解，主要途径是酶解（enzymatic hydrolysis）（图 1.30）和酸解（acid hydrolysis）。

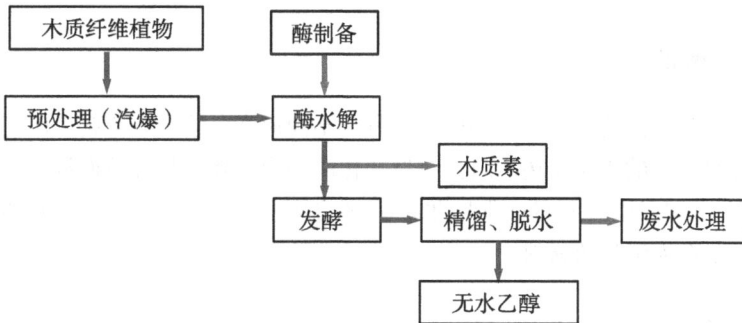

图 1.30　木材加工燃料乙醇的工序流程

在气化剂的作用下，木质原料通过高温热化学反应转化为可燃气体（combustible gas），再将净化后的可燃气体送入内燃机直接发电就可以获得电能（图 1.31）。2000 年海南三亚木材厂成立了首家木屑气化发电厂，该发电厂以木材废料为原料，取得了很好的节能环保效益。

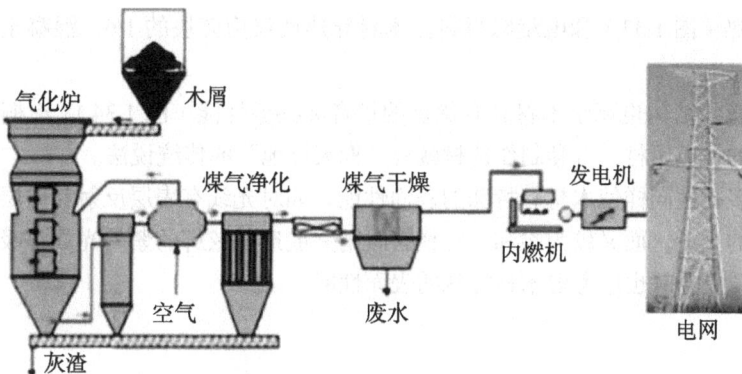

图 1.31　木屑气化发电的工序流程

1.2.7　木材制浆造纸（pulping and papermaking）

纸是人民文化生活和日常生活必不可少的物品。木材制浆造纸是森林资源综合利用的重要途径之一，也是提高木材高附加值的一种重要的加工方法。

木材制浆是利用机械方法或化学方法，或者二者结合的方法，把木材原料的纤维离解开来的过程。制浆方法分3种：机械法、化学法、化学机械法。

木材制浆造纸的特点是：

①原料来源广，大径材、小径材、薪材，以及采伐、造材、加工剩余物，把皮去掉后，可全部利用。

②碱容易回收，对环境污染少。

1.3　木材的特性

1.3.1　生物学特性

木材以年轮为主体的花纹，在不同切面具有不同形态，给人以流畅、轻松、协调和高雅的感觉。宽窄不一的年轮是树木对大自然变化的响应，随生长环境而波动，它与生物固有波动相吻合[①]，最易引起人们心理上的共鸣，也能给人有安全感的韵律。所以，木材是与人类最亲近、最有人情味的材料。

1.3.2　多孔性（porosity）

木材由树木生长产生的细胞所组成。它与树木中活细胞的主要区别是：它已成为木质化的死细胞。死细胞由细胞壁（cell wall）和细胞腔（cell lumen）组成，因此木材是含大量孔隙的多孔性材料。这种天然形成中空而有较大强度的生物材料，具有较高的强重比（strength-weight ratio）、较好的刚性和吸收冲击荷载等优良的力学性质（图1.32）。

木材的多孔性，使木材密度较低，细胞腔中存在空气，空气是热和电的不良导体，所以木材是隔热（图1.33）和电绝缘材料。木材导热性仅为砖块的1/6，混凝土的1/15，钢材的1/390。

多孔的管状结构也赋予木材具有优良的扩音和共振性能（图1.34）。泡桐等木材可用于许多乐器的音板用材，古琴制作选材就有"桐天梓地"的传统说法。

木材的多孔性同样使木材有特殊的表面性能，如对光线有表层反射和内层反射，使木材具有一定的光泽，而又较为柔和；对伤害眼睛并能造成眩辉的紫外光具有吸收作用。此外，在油漆和胶合时也应考虑木材特殊的表面性能。

---

① 木材不同部位的木纹图案呈现着"涨落"周期式变化规律（谱分布形式），暗合人体生物涨落节律（如 α 脑波的涨落、心动周期的变化也为谱分布形式）。

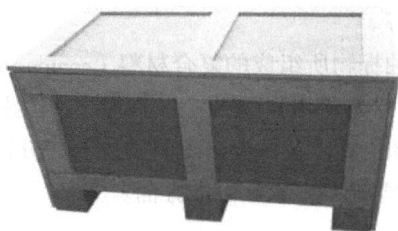

图 1.32　精密仪器木材包装箱

图 1.33　保温瓶木塞

（a）钢琴

（b）古筝

图 1.34　乐器

多孔性也增加了木材的比面积，木质活性炭[①]就是利用了木材的多孔特性（图 1.35）。多孔性也增加了木材吸湿性，造成木材尺寸不稳定。多孔性使木材有与其他材料不同的气体和液体渗透性，对木材干燥、防腐、改性等处理时，选择工艺应充分考虑渗透性。

图 1.35　木质活性炭

1.3.3　各向异性（anisotropy）[②]

树木生长是由树皮和木质部之间的形成层分生而逐年加粗，形成近似轮层圆柱体。从髓心向树皮有横向排列的射线细胞，形成辐射状的木射线组织，因此木材为近似圆柱对称体，有纵向（longitudinal）、径向（radial）和弦向（tangential）3 个基本方向。不同方向上木材的水分传导、干缩（shrinking）、热、电、声及强度等性质均有所不同。

[①]　木质粉状活性炭是以木屑为原料精制而成，外观为黑色细微粉末状，无毒、无味，具有比表面积大，吸附能力强，适用于净水行业，在有机物溶剂的脱色、精制、提纯和污水处理方面被广泛使用。

[②]　各向异性指某一物体的全部或部分化学、物理等性质随着方向的改变而有所变化，在不同的方向上呈现差异的性质。

1.3.4 耐久性

木材是由纤维素、半纤维素、木质素和抽提物所组成的复合材料（composite）。木材的这些组分均含有亲水的羟基，具有吸湿性，会影响木材的尺寸稳定性；一定湿度下可成为菌、虫的营养物质，造成木材变色、腐朽和虫蛀。在常温下木材的主要成分性质稳定，与大多数化学药剂不起化学反应，能用于抗化学药剂要求较高的方面。

1.3.5 可再生性

木材是生物材料，属于可再生资源材料，同时又是新能源材料。按世界现有森林蓄积量，每年至少能增加 30 亿 m^3，与地下资源相比具有明显的优越性。矿物资源按现阶段使用水平，石油在 21 世纪将枯竭，银、水银和锌仅够开采 20 年；锡和铅 40~50 年；铜和镍 60~80 年；铁和锰 160~170 年。只要加强森林抚育，木材就能成为取之不尽、用之不竭的再生资源和能源材料。

▶ **思考与训练**

1. 我国森林资源的特点是什么？
2. 森林的生态功能有哪些？
3. 木材的应用领域有哪些？
4. 木材的特性有哪些？

（1）了解树木生长与木材形成过程；

（2）了解植物分类知识；

（3）了解木材的来源；

（4）了解木材的命名规则。

▶ 本章描述

植物怎么分类？有什么规定？木材主要来源于哪些植物？木材来源于树木，那么树木如何长高、长粗？树木的哪一部分能够产生木材？木材能够生产哪些产品？木材该如何命名？对上述问题的了解，是科学利用木材的基础。

木材来源于复杂的生物体——树木。通过**无性繁殖**（vegetative propagation）或从受精卵变成胚胎的有性繁殖，树木长成自然界最大的生物体之一。和人类一样，树木幼时也很脆弱。在适当的环境下，树木才能茁壮成长。

就木材来源而言，木材通常是指高度达到 4~6m 的乔木的**树干**（trunk），而且往往是树干中的**木质部**（xylem）。具体而言，**木材**（wood）是由**形成层**（cambium）分生的次生**木质部**（secondary xylem），是以输导水分为中心的几种细胞的集合。

2.1 植物分类与木材来源

全世界的植物大约有 50 万种。因此需要把纷繁复杂的植物界分门别类，并按系统排列起来，以便于人们认识和利用植物。

植物分类常采用的方法是**恩格勒（Engler）系统**的自然分类法。此分类法是根据植物的花、果、叶的主要形态特征进行分类的。常用的植物分类单位是界、门、纲、目、科、属、种。种是最基本的分类单位，具有杂交不育的特征。各分类单位根据需要再分成亚级，如亚门、亚纲、亚目、亚科、亚属。种下面又分为亚种、变种和变型。亲缘相近的种集合为属，亲缘相近的属集合成科，亲缘相近的科集合成目，亲缘相近的目集合成纲，以此类推。

以柏木和刺槐为例，说明植物分类如下：

界	植物界	界	植物界
门	种子植物门	门	种子植物门
亚门	裸子植物亚门	亚门	被子植物亚门
纲	球果（松杉）纲	纲	双子叶植物纲
目	松杉目	目	蔷薇目
科	柏科	科	蝶形花科
属	柏属	属	刺槐属
种	柏木	种	刺槐

植物界分为四大门：**藻类植物、苔藓植物、蕨类植物和种子植物**（图 2.1）。

藻类植物是一类比较原始、古老的低等生物。藻类的构造简单，没有根、茎、叶的分化，多为单细胞、群体或多细胞的叶状体。如小球藻是单细胞，团藻属于群体，海带呈叶状体。藻类含叶绿素等光合色素，能进行光合作用，属自养型生物。

苔藓植物是一类生活史中以配子体占优势的小型绿色植物，结构简单，无维管组织，没有茎、叶、根的分化，我们平时所见到茎、叶和根，严格上应该叫拟茎、拟叶和拟根。

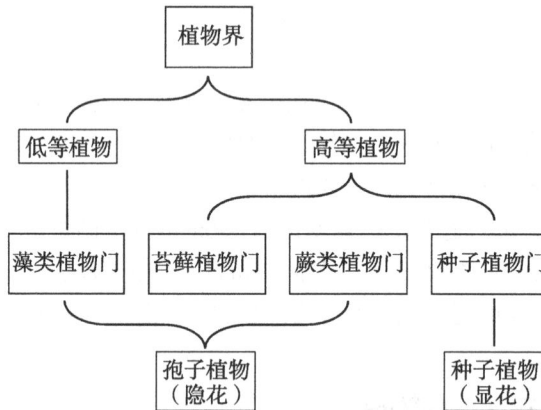

图 2.1　植物分类

蕨类植物又称羊齿植物，是一群进化水平最高的孢子植物。生活史为孢子体发达的异形世代交替。孢子体有根、茎、叶的分化，有较原始的维管组织。配子体微小，绿色自养或与真菌共生，有根、茎、叶的分化。

种子植物分布于世界各地，是植物界最高等的类群，所有的种子植物都有两个基本特征，就是体内有维管组织（韧皮部和木质部）、能产生种子并繁殖。

种子植物是现今地球上种类最多、形态构造最复杂的一群植物，全世界有 20 万 ~25

万种，中国有 3 万种，是和人类经济生活最密切的一类植物。种子植物可分为**裸子植物**（gymnosperm）和**被子植物**（angiosperm）。

裸子植物为多年生木本植物，大多为单轴分枝的高大乔木，少为灌木，稀为藤本。

被子植物是当今世界植物界中最进化、种类最多、分布最广、适应性最强的类群。被子植物可以分为**单子叶植物**和**双子叶植物**。单子叶植物和双子叶植物都可以分为**木本植物**①（woody plant）和**草本植物**②（herb）（图 2.2）。

（a）木本植物　　　　　　　　　　　　　（b）草本植物

图 2.2　木本植物和草本植物

树木属种子植物类，是木本植物的总称，包括多年生的高大乔木、低矮丛生的灌木和缠绕他物的藤本植物（图 2.3）。而**木材主要来源于种子植物中裸子植物的乔木和被子植物中双子叶植物的乔木**（图 2.4）。

在裸子植物中，银杏纲是裸子植物中古老原始的一个纲，仅存银杏目。该目仅含 1 科 1 属 1 种。银杏又称白果树、公孙树、鸭掌树。木材材质优良，结构细，宜作翻砂模型及印染机滚筒、绘图板、雕刻、工艺品和室内装饰等用材。松杉纲有 4 目 7 科，4 目为：三

---

① 木本植物是指根和茎因增粗生长形成大量的木质部，而细胞壁也多数木质化的坚固的植物。植物体木质部发达，茎坚硬，多年生。

② 草本植物是指茎内的木质部不发达，含木质化细胞少，支持力弱的植物。草本植物体形一般都很矮小，寿命较短，茎干软弱，多数在生长季节终了时地上部分或整株植物体死亡。根据完成整个生活史的年限长短，分为一年生、二年生和多年生草本植物。

（a）乔木  （b）灌木  （c）藤本

图 2.3　木本植物

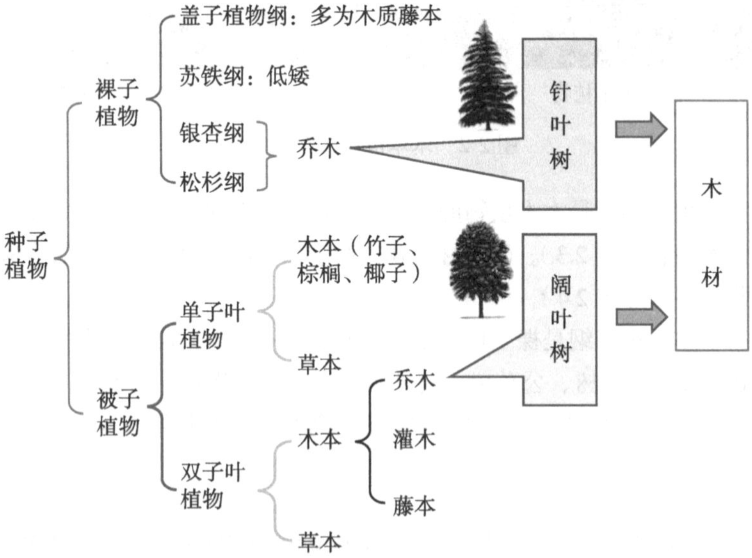

图 2.4　木材的来源

尖杉目、松杉目、罗汉松目、红豆杉目；7 科为：三尖杉科①、南洋杉科②、松科③、杉科④、柏科⑤、罗汉松科⑥、红豆杉科⑦。比起针叶树，阔叶树种类更多⑧。

若从木材产品的角度进行分类，木材有立木、伐倒木、原条、原木、锯材、木质板材、重型木基结构材之分（图 2.5）。

◎ **立木**（standing tree）：林地中生长着的林木。

◎ **伐倒木**（felling tree）：伐倒后的林木。

◎ **原条**（pole）：伐倒木去枝。

◎ **原木**（log）：原条按一定长度规格截断后。

◎ **锯材**（lumber）：原木纵向锯解形成。锯材有方材和板材之分，宽度 / 厚度小于 2 的锯材称为方材，宽度 / 厚度大于 2 的锯材称为板材。

◎ **木质板材**（wood-based panel）：通过一定生产工艺对木材及其剩余物进行深层次加工，使其成为板材，传统的木质板材为胶合板、纤维板和刨花板。

立木　　伐倒木　　原条　　原木

方材

板材

木质板材

重型木基结构材

图 2.5　木材产品

---

① 三尖杉科仅有 1 属：三尖杉属。

② 南洋杉科有 2 属：贝壳杉属和南洋杉属。

③ 松科有 3 个亚科：冷杉亚科（冷杉属、银杉属、油杉属、云杉属、黄杉属、铁杉属）、落叶松亚科（雪松属、落叶松属、金钱松属）和松亚科（松属）。

④ 杉科有 9 属：柳杉属、杉木属、水松属、水杉属、金松属、北美红杉属、巨杉属、台湾杉属、落羽杉属。

⑤ 柏科有 3 个亚科：柏木亚科（扁柏属、柏木属、福建柏属）、圆柏亚科（刺柏属、圆柏属）、侧柏亚科（翠柏属、侧柏属、崖柏属、罗汉柏属）。

⑥ 罗汉松科有 2 属：陆均松属、罗汉松属。

⑦ 红豆杉科 4 属：穗花杉属、白豆杉属、红豆杉属、榧树属。

⑧ 阔叶树的植物分类学比针叶树的更复杂，开花植物有三四百个科，但是只有少数（大约 80 科）是纯粹的树木。

◎ **重型木基结构材**（mass timber product）：使用胶黏剂或钉子或木销，把锯材（或
称规格材）用一定的方式层合在一起，制造成尺寸较大的板材，属于新型木质板
材，主要用于木结构建筑。

2.2　树木生长与木材形成

　　树木的生命开始于一颗种子（seed），而后经历一个生命周期（life cycle）。有些树能
够存活几个世纪。有些树能够长得十分高大。在美国加利福尼亚州有一株古老的北美红杉
树，即树龄已超过 2400 年的"谢尔曼将军树"（General Sherman Tree）（图 2.6），是现今存
活的最大生物体。树高超过 80m，胸径达 11.1m，单株材积超过 900m³。

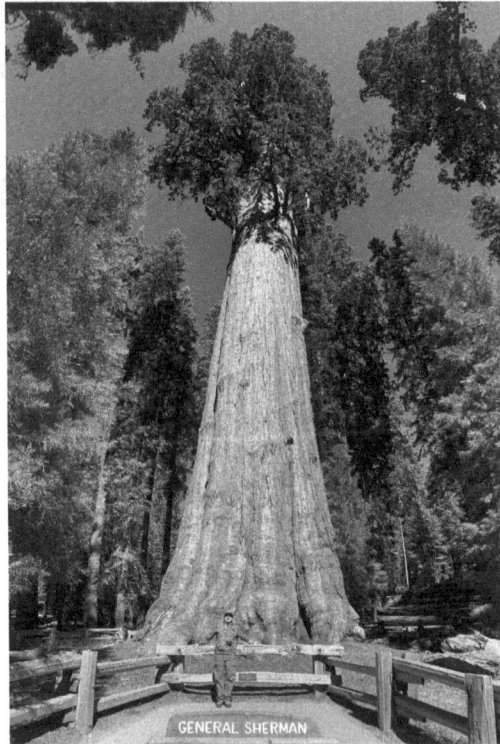

图 2.6　美国红杉"谢尔曼将军树"

　　树木的木质部被树皮覆盖。树皮由内层树皮（韧皮部）（inner bark / phloem）和外层树皮
（外层保护层）（outer bark）组成。随着树木的生长，会产生新的木材，主茎（stem / trunk）
和主干（branch）的直径随之增加。树皮也会不断增加，同时也会不断开裂和脱落。

　　和其他绿色植物一样，树木有多种栽植方式。如种子成熟后从树上掉落在地上，然
后随着风吹在动物（如鸟儿）的搬运四处散播，或者种子也可以在苗圃的培养皿中培育

获得。在合适的条件下，种子发芽，胚芽[①]（plumule）形成茎；而胚根继续在土壤中生长。萌芽进一步长出树根和叶子，形成幼苗。随着树木逐渐强壮，幼苗形成幼树（图 2.7）。幼树继续长大，最后长成枝叶繁茂、根系发达的高大乔木。

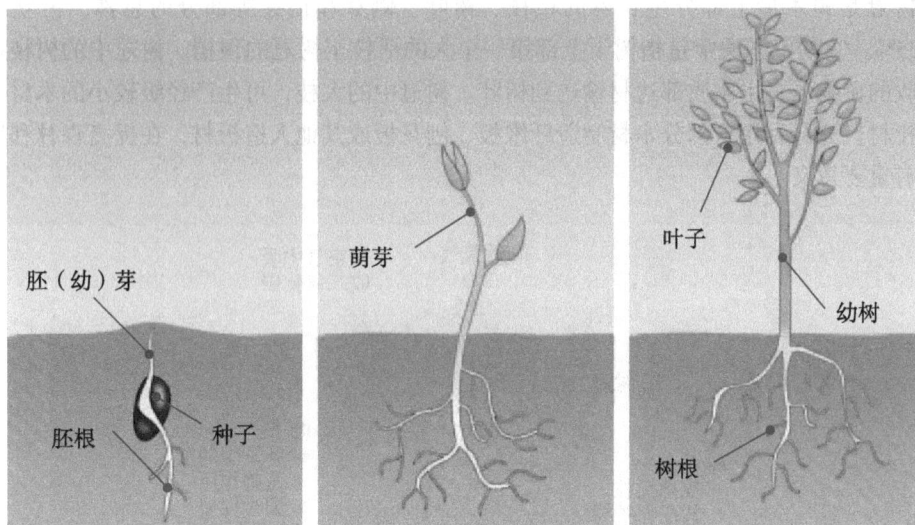

胚（幼）芽

萌芽

叶子

幼树

胚根　种子

树根

图 2.7　幼树的生长

伴随着幼树长成高大乔木，叶子借助叶绿素（chlorophyll）进行光合作用[②]（photosynthesis）。光合作用需要的水来自土壤，需要的二氧化碳来自大气，需要的光来自太阳。纵观全树，树木由树根（root）、树干和树冠（crown）三大部分组成（图 2.8）。

树根是树木的地下部分，由主根、侧根和须根组成，占立木[③]（standing tree）总体积的5%~25%。主根的功能是支持树体，将强大的树冠和树干稳固地固定在土壤中，保证树木的正常生长。侧根和须根则主要是从土壤中吸收水分和矿物质，供树冠中的叶片进行光合作用。

树干是树冠和树根之间的直立部分，是树木的主体，占立木体积的 50%~90%。在活树（living tree）中，树干具有输导、储存和支撑的重要功能。木质部的边材（sapwood）把树根吸收的水分和矿物质上行输送到树冠。再把树冠制造出来的有机养料部分通过树皮的韧皮部下行输送到树木全体，部分储存在树干内。木射线（wood ray）为树木水平方向

---

① 胚是构成种子最重要的部分，它由胚芽、胚根、胚轴和子叶 4 部分组成。种子萌发后，胚根、胚芽和胚轴分别形成植物体的根、茎、叶及其过渡区，因而胚是植物新个体的原始体。胚芽位于胚轴的顶端，突破种子的皮后发育成叶和茎。

② 光合作用通常是指绿色植物（包括藻类）吸收光能，把二氧化碳和水合成富能有机物，同时释放氧气的过程。其主要包括光反应、暗反应两个阶段，涉及光吸收、电子传递、光合磷酸化、碳同化等重要反应步骤，对实现自然界的能量转换、维持大气的碳氧平衡具有重要意义。

③ 立木指形成森林主要部分的树木的总和，亦指林地上未伐倒的活着的树木。

上运输营养物质提供了通道。射线也有储存碳水化合物①的功能。在休眠期②过后，射线可以作为储存物质从树中心（也称髓心，pith）向外水平运输的通道，它也是旋切单板干燥中水分蒸发的主要通道。

树冠是树木最上部分生长着的枝桠、树叶、侧芽和顶芽等部分的总称，占立木的5%~25%。它的范围通常是指树干上部第一个大的活枝至树冠的顶梢。树冠中的树枝把根部吸收的矿物质，由木质部边材输送到树叶。树冠中的大枝，可生产径级较小的木材，统称枝桠材。充分利用这部分木材制造纤维板、刨花板或其他人造板材，在提高森林利用率上具有重要意义。

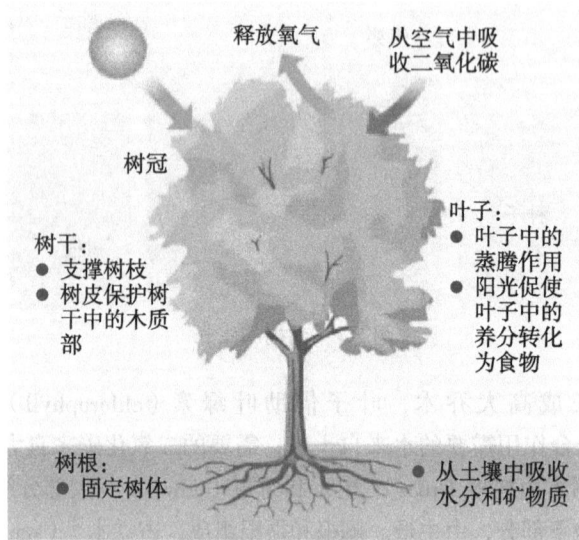

图 2.8　树木的组成

木质部和韧皮部③之间的一个薄层会产生新的木质部和韧皮部组织，这一层被称为形成层。形成层完全包裹着小枝、树枝、树干和根，这意味着一个生长季节会在整棵树上形成一个新的连续的木材层。树干是木材的主要来源，它由 4 个部分组成，即树皮④、形成层、木质部（边材和心材）及髓（图 2.9）。髓与初生木质部合称髓心。

树木生长是指树木在生长发育过程中，通过细胞分裂和扩大，使树木的形体和重量不断增加的过程。树木是多年生植物，它的一生要经过幼年期、成熟期和过熟期，直至衰老死亡。

---

① 碳水化合物（carbohydrate）是由碳、氢和氧 3 种元素组成，自然界存在最多、具有广谱化学结构和生物功能的有机化合物。

② 休眠期是植物体或其器官在发育的过程中，生长和代谢出现暂时停顿的时期。

③ 被子植物的韧皮部由筛管和伴胞、筛分子韧皮纤维和韧皮薄壁细胞等组成，位于树皮（狭义）和形成层之间。其可分为初生韧皮部和次生韧皮部两种。

④ 狭义的树皮包括 3 层：木栓、木栓形成层和栓内层，以及外部的各种死组织。广义的树皮还包括韧皮部，由内到外包括韧皮部、皮层和多次形成累积的周皮以及木栓层以外的一切死组织。

内层树皮　外层树皮

髓

形成层

心材　边材

木质部

图 2.9　树干的组成部分

2.2.1　高生长（height growth）

图 2.7 中的幼树根系发育良好，树冠为 1~2 年树龄的典型树冠。在早春季节，树木开始生长，通过细胞的形成和增大，每根枝条顶端的芽随着组织的膨胀而膨胀。这些细胞反复分裂形成新细胞的区域称为**分生组织区**（meristematic region）。每个分枝的顶端都有相似的芽。主茎顶端的分生组织区具有特殊的意义，因为它在一定程度上控制着分枝和嫩枝的发育，因而它被称为**顶端分生组织**（apical meristem）。

顶端分生组织的细胞分裂作用是延长主茎。在这个位置的细胞增长产生了新细胞，导致树木增高，这就是树木的高生长（图 2.10）。显然，高生长是树木根和茎主轴生长点（即顶端分生组织）分生活动的结果。生长点在生成新的细胞后，树茎就产生高度伸展，使树木长高。这种生长是把新长成的细胞留置在下方，而把生长点向上抬高。树木因生长点的分生作用而引起的这种高生长，称为初生长。由生长点所形成的组织，称为初生组织。因为树木是从顶端而不是从底部生长的，所以如果在地面以上 2m 的树体上钉钉子，不管树长到什么高度，钉子离地高度几乎不变。

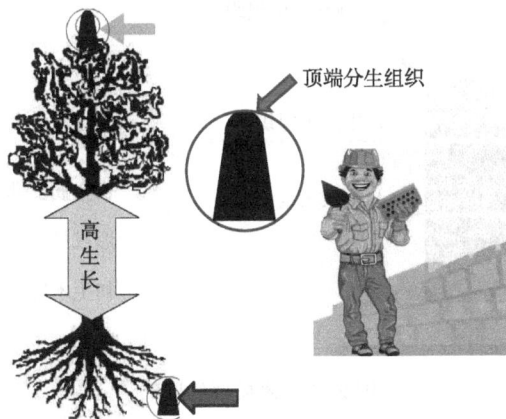

顶端分生组织

高生长

图 2.10　树木高生长

可以通过工人砌砖墙形象类比树木高生长过程，顶端分生组织相当于泥瓦匠，泥瓦匠位置不断抬高且始终位于墙体顶部（图 2.10）。

2.2.2 直径生长（diameter growth）

树木的直径生长是木质部和韧皮部不断增加的结果，它是由形成层原始细胞进行弦向**平周分裂**来完成的。在向髓心方向增加的细胞远较向外增加的细胞多得多，久而久之，树木的直径便不断增大，形成层也随之外移。植物学上，形成层被称为侧生分生组织，由它分生出来的组织叫次生组织。次生组织包括由形成层所形成的次生木质部（secondary xylem）和次生韧皮部（secondary phloem）（图 2.11）。树木因形成层的分生作用引起的直径增粗生长，称为次生长。随着内部的变化，形成层本身也要进行细胞分裂，这种自身增生的分裂，称为**垂周分裂**。

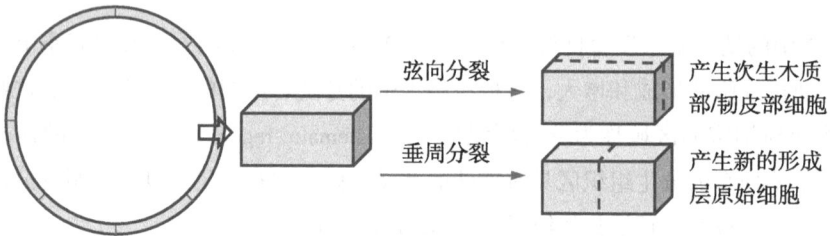

图 2.11　形成层的分裂

木质部位于形成层和髓之间，是树干的主要部分。根据细胞组织的来源，木质部可分为初生木质部和次生木质部。顶端分生组织形成初生分生组织，初生分生组织失去分裂能力形成初生永久组织，初生永久组织包括：表皮、皮层、初生韧皮部、形成层、初生木质部和髓（图 2.12）。树木完成高生长后，初生永久组织的形成层再进一步分生，进行直径生长，产生了次生木质部。

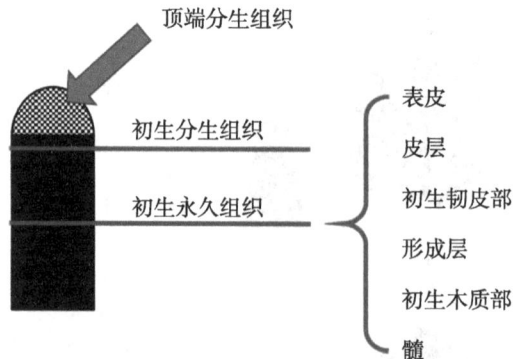

图 2.12　木质部的形成

2.3　木材的命名

木材和其他物种一样，各有其名称，如松木、柏木和杨木等。一种木材在不同的地方有不同的名称；有时在同一个地方也有几种或多种名称，这种现象叫同物异名。还有一种情况是异物同名，如酸枣在南方各地是指槭树科的一种植物，而在北方各地是指鼠李科的一种植物。又如松木的一般概念是指松属木材的多种或一种木材；同时也指除了柏木、杉木以外的几乎全部的针叶树材，有时甚至用作针叶树材的同义词，这就出现了木材名称的混杂现象。因此，有必要了解木材名称，以便更好地识别木材及促进木材贸易。

2.3.1　学名

每种植物在全世界通用的名称为学名。学名是由拉丁文或拉丁化的其他外文组成。

每一学名包括属名和种名，即采用"双命名法"，种名后附命名人姓氏。属名的首字母大写，属名和种名一般多采用斜体书写。为简便起见，常常舍去命名人姓氏。如马尾松的学名为 *Pinus massoniana*。

学名在国际交流和科学鉴定等方面有着实际意义。学名固然科学，但由于文字障碍和木材树种过于繁杂，在应用中仅凭肉眼不易确定到种，故在木材生产、贸易和使用等领域受到一定的限制。再者，宏观特征和材质差别不大的树种，使用价值近乎相同，没有必要区分到种。如从北美进口的云杉—松—冷杉（spruce-pine-fir）规格材，就把这三类树种放在一起销售和使用。

2.3.2　商品名

用于市场交易的木材称为商品材。通常以树木分类中的属为基础，材质为主要依据，将宏观构造相似，木材材质差异不大和现场难以区别的商品材树种归类，并以属名的树种标准名称作为木材的商品名。一个商品名有的包括全属的树种，如泡桐属的各树种，其商品名均为泡桐；有的包括属内部分树种，如松木（或硬松）为松属中马尾松、樟子松和油松等树种的商品名；有的则包括不同属的树种，如白青冈包括青冈栎属中的青冈栎和麻栎属中的乌冈栎等。

红木不是一个树种，而是紫檀属、黄檀属、柿属、崖豆属及决明属树种满足一定的构造特征、密度和材色的一类木材集群。我国国家标准《红木》（GB/T 18107—2017）列出了 29 个称为红木的树种。

2.3.3　俗名

木材俗名是木材的通俗叫法，具有地方性。由于各地的取名习惯和语言差异，所使用的木材名称不尽相同。如市场上所谓的"榉木"，实际上指的是壳斗科水青冈（山毛榉）属（*Fagus* spp.）的木材，而真正的榉木则是榆科榉属树种（*Zelkova* spp.）。

可见，木材俗名的不规范性必然带来同物异名或同名异物的混乱，给木材流通等带来障碍。

▶ 思考与训练

1. 植物的科学命名法是采用双名法，该命名法是由（ ）创立的。

A. 牛顿　　 B. 林奈　　 C. 伽利略　　 D. 列文·胡可

2. 乔木树种属于（ ）植物门。

A. 种子　　 B. 苔藓　　 C. 蕨类　　 D. 藻类

3. 树木的生长是指树木的高生长吗？

4. 植物分类的等级中最基本的分类单元是_____。

5. 试述银杏和鹅掌楸的植物分类系统层级。

6. 如图所示，在盛有被稀释的红墨水中，插入一根有几片树叶的木本植物枝条，放置在阳光下。讨论，茎的哪一部分先被染红？

第❸章
木材构造

▷ **学习目标**

（1）了解木材的主要宏观构造特征；

（2）了解木材的主要显微构造特性；

（3）了解木材的识别方法；

（4）了解国产和进口主要商品材及特点。

▷ **本章描述**

不同树种的木材各有特点，如何识别和鉴定？有没有共性？在肉眼看不见的地方，显微镜下的木材构造呈现什么状态？了解木材构造，才能因材适用，物尽其用。

木材是一种天然材料，由中空、细长、纺锤形的厚壁细胞沿着树干方向和近等径薄壁细胞沿着径向排列组成。这些细胞及其排列会影响木材的强度、刚度、干缩湿胀和表面纹理等性质。研究木材构造的目的，在于揭示不同树种间木材构造的共性和差异性，从而提升人们认识木材和利用木材的水平。

木材构造特征依据采用的工具和放大倍数而分为 3 个层次：

◎ 用肉眼或放大镜所观察到的木材构造特征，放大倍数在 5~10 倍，为宏观构造特征（macroscopic / gross features）；

◎ 借助于普通光学显微镜观察到的木材构造特征，放大倍数在 200 倍以下，为显微构造特征（microscopic / minute features）；

◎ 近代应用 X 射线和电子显微镜显示出来的木材细胞壁的构造特征，放大倍数在 1000 倍以上，为超微构造特征[①]（ultramicroscopic features）。

上述方法在木材科学研究及木材合理加工与增值利用中都有重要作用。本章主要从宏观和显微 2 个层面介绍木材构造。

---

① 细胞壁结构将在任务 5 中介绍。

3.1 木材的主要宏观构造

木材的宏观构造也被称为木材粗视（gross）构造，是指用肉眼或借助 10 倍放大镜所能观察到的木材构造特征。

3.1.1 木材三方向

根据树木的生长方向和木材细胞的排列方向，可以把木材的方向定义如下（图 3.1）：

◎ **轴向**（longitudinal direction），指树干的长轴方向，也就是木材的顺纹方向。

◎ **径向**（radial direction），指生长轮（年轮）的直径方向。

◎ **弦向**（tangential direction)，指生长轮（年轮）的切线方向，与直径方向垂直。

其中径向和弦向又统称为横向（transverse direction）。

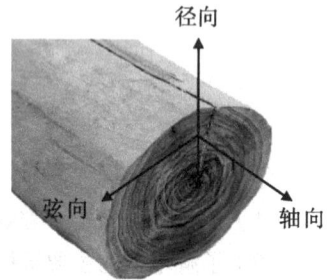

图 3.1　木材三方向

3.1.2 木材三切面

横切面（cross-sectional surface）：与树干长轴相垂直的切面，亦称端面或横截面，如图 3.2（a）所示。在这个切面上，木材细胞间的相互联系都能够清楚地反映出来。年轮或生长轮在横切面上呈同心圆状。在应用上，由于这个切面硬度大、耐摩擦，故可以用作砧板和铺路木板。

径切面（radial surface）：顺着树干长轴方向，通过髓心与木射线平行或与生长轮相垂直的纵切面，如图 3.2（b）所示。在这个切面上，年轮呈条状，相互平行，且与木射线互相垂直。由径切而成的板材，收缩小，不易翘曲，适用于地板、木尺和乐器用材的共鸣板。

弦切面（tangential surface）：顺着树干长轴方向，不通过髓心，与木射线垂直或与生长轮相平行的纵切面，如图 3.2（c）所示。年轮或生长轮在弦切面上呈抛物线状或倒 "V" 字形花纹。这个切面上的花纹非常漂亮，因而拥有弦切面的木材可用于制造高档家具。

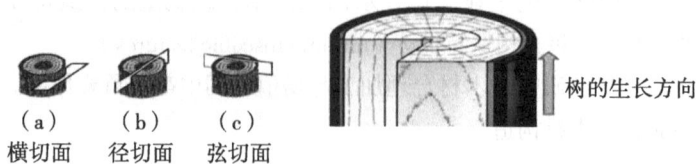

（a）横切面　（b）径切面　（c）弦切面

图 3.2　木材三切面

3.1.3 生长轮（年轮）

生长轮（growth ring）即树木在一个生长周期内，形成层向内分生的一层次生木质部，在横切面上呈现为围绕着髓心的同心圆（图 3.3）。在温带、寒带及亚热带地区，树木一年内仅生长一层木质部，因此生长轮又称为**年轮**（annual ring）。而在热带地区，部分树木生长季节仅与雨季和旱季的交替有关，一年内会形成多层木质层。生长轮数目的多少，与木材性质有一定的关系。在木材利用上，是以横切面上垂直年轮方向 1cm 内的数目来估计木材的物理力学性质的，年轮数目越多，木材的密度和强度也越大。

3.1.4 木材心边材

在木质部中，靠近树皮（通常颜色较浅）的外环部分称为**边材**（sapwood），髓心与边材之间的木质部（通常颜色较深）称为**心材**（heartwood）（图 3.4）。树木在幼龄时期全部由边材构成，随后，部分边材逐渐转变为心材。边材转变为心材的过程是一个复杂的化学变化和生物化学变化。

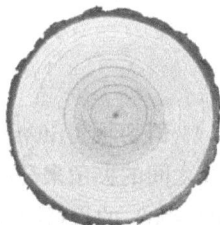

图 3.3 木材生长轮　　　图 3.4 木材心材与边材

3.1.5 木材早晚材

每一个年轮是由两部分组成。靠近髓心一侧，是树木每年生长季节早期（春季或夏季）形成的一部分，称为**早材**（earlywood）。而靠近树皮一侧，是树木每年生长后期形成的一部分木材，称为**晚材**（latewood）（图 3.5）。

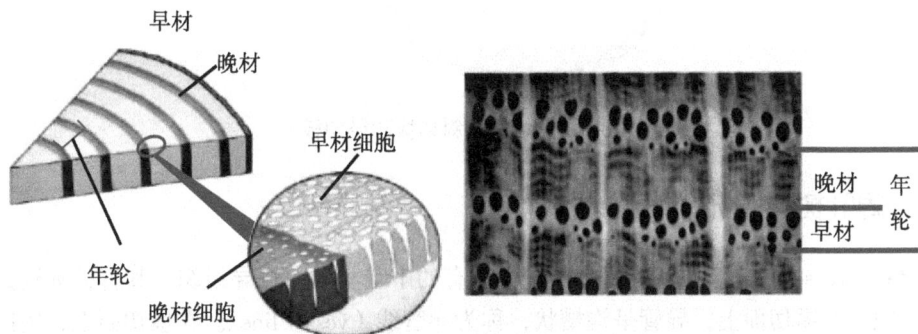

图 3.5 木材早材和晚材

对于温带、寒带和亚热带生长的树木来说，每年春夏雨水较多，气温高，水分、养分充足，形成层细胞分裂速度快，所形成的早材细胞壁薄，形体较大，材质较疏松，颜色较浅。秋冬雨水少，气温低，水分、养分不足，形成层细胞分裂速度慢，所形成的晚材细胞壁厚，形体较小，材质较致密，颜色较深。

由于早晚材结构不同，在两个年轮交界处的组织有明显差异，明显地衬托出一条界线，称为**轮界线**（ring boundary）。它的明显与否，称为年轮明显度。针叶树材的早晚材差异显著，故其轮界线较阔叶树材的明显。

针叶树材的材性通常可以通过计算**晚材率**来评估，晚材率越高，其密度和强度也越高。每一年轮中的晚材率 P 可按照下式计算：

$$P = \frac{b}{a} \times 100\%$$

式中：b——年轮中晚材的宽度，mm；

　　　a——年轮总宽度，mm。

晚材率是衡量木材强度大小的一个重要标志。晚材率越高，密度越大，木材强度越大。晚材率在横切面的径向，自髓心向外逐渐增加，但达到最大限度后开始降低。

3.1.6　径切板、弦切板与斜切板

木材加工中通常所说的**弦切板**（plain-sawn board）、**斜切板**（rift-sawn board）和**径切板**（quarter-sawn board）是以板厚中心线与生长轮切线之间的夹角来分的，其夹角 0~30° 的板材称为弦切板，30°~60° 的板材称为斜切板，60°~90° 的板材称为径切板（图3.6）。

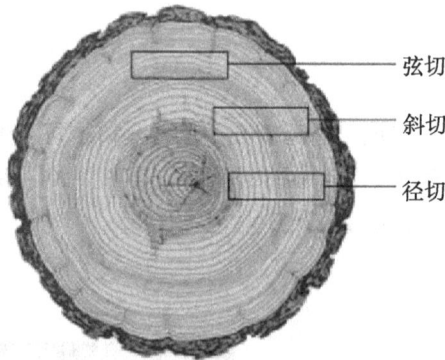

弦切

斜切

径切

图 3.6　弦切板、斜切板和径切板

3.1.7　阔叶树材管孔

导管（vessel）是绝大多数阔叶树材所具有的中空状轴向输导组织，其直径远远大于其他细胞。在纵切面上，导管呈沟槽状，称为导管线（vessel line）。在横切面上，用肉眼可以看到许多小孔，称为**管孔**（pore）。所以，阔叶树材称为**有孔材**（porous wood）。针叶树材无导管，在横切面上肉眼看不到管孔，所以针叶树材称为**无孔材**（nonporous wood）。

根据管孔在横切面上一个生长轮内的分布和大小情况，一般可以将阔叶树材划分为**环孔材**（ring-porous wood）、**半环孔材**（semi ring-porous wood）和**散孔材**（diffuse-porous wood）（图 3.7）。

环孔材指的是在一个生长轮内，早材管孔比晚材管孔显著大些，并沿生长轮方向呈环状排列成一至数列（如白蜡木）。

半环孔材也称半散孔材，指的是在一个生长轮内，管孔的排列介于环孔材与散孔材之间，早材开始部分的管孔较晚材末端部分的管孔显著为大，且在多数情况下沿生长轮成稀疏状排列。但早材管孔逐渐向晚材部分变小，早晚材之间的过渡无明显的界线，生长轮内管孔分布比较均匀（如黑胡桃）。

散孔材是指在一个生长轮内，早晚材管孔的大小无显著的区别，分布均匀或比较均匀（如桦木）。

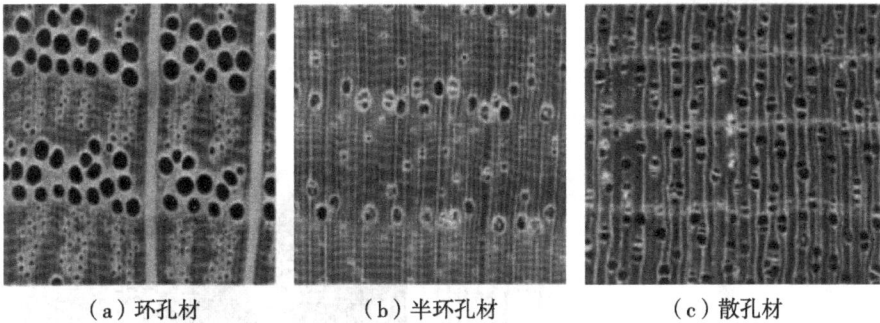

（a）环孔材　　　　　　（b）半环孔材　　　　　　（c）散孔材

图 3.7　管孔分布

3.1.8　木射线

木材横切面上可以看到一些颜色较浅或略带光泽的线条，它们沿着半径方向呈辐射状穿过年轮，称为**木射线**（wood ray）（图 3.8）。木射线是树木的横向组织，来自于射线原始细胞，起横向运输和储存养料的作用。

图 3.8　木射线

木材干燥时，最易沿木射线开裂。但具有宽木射线的木材，如栎木、水青冈等，其径切面常呈现出银光纹理，构成美丽花纹，适于做家具及细木工的材料。

3.1.9　胞间道

胞间道（intercellular canal）是由分泌细胞环绕而成的长度不定的管状细胞间隙。针叶树材中储藏树脂的胞间道称为树脂道（resin canal）；阔叶树材中储藏树胶的胞间道称为树胶道 (gum canal)。松科 6 属（即松、落叶松、云杉、黄杉、银杉、油杉）木材具有正常树脂道。图 3.9 所示为松科松属马尾松的树脂道及树脂产品。

（a）马尾松　　　　　　　　（b）树脂道及树脂产品

图 3.9　马尾松树脂道及树脂产品

3.1.10　木材纹理、肌理与花纹

木材纹理（grain）是指木材体内轴向分子（如木纤维、管胞、导管）排列方向的表现形式。

根据木材纹理方向通常分为 3 种情况（图 3.10）：

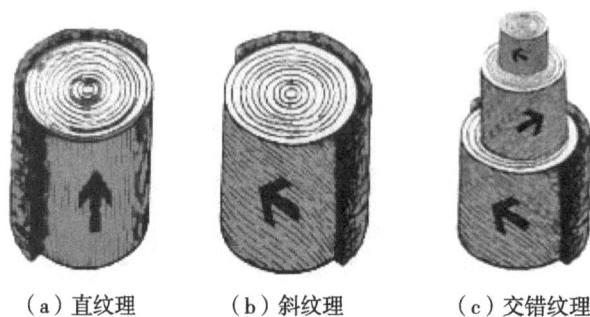

（a）直纹理　　　　　（b）斜纹理　　　　　（c）交错纹理

图 3.10　木材纹理

◎ 排列方向与树干基本平行的纹理称为直纹理（straight grain），如红松、杉木和榆木等；

◎ 排列方向与树干不平行，呈一定角度的倾斜称为斜纹理（spiral grain），如圆柏、枫香和香樟等；

◎ 排列方向错乱，左螺旋纹理与右螺旋纹理分层交错缠绕的纹理称为交错纹理（interlocked grain），如海棠木、大叶桉等。

木材肌理（texture）指木材细胞大小及差异的程度，如粗肌理、细肌理。

木材花纹（figure）是以木材纹理和肌理为主体，木材表面因年轮、木射线、轴向薄壁组织、木节、树瘤、材色以及锯切方向不同等产生的美丽图案。

木材花纹与木材构造密切相关，不仅有利于木材识别，而且经适当利用，还可以提高木材利用价值。由于木材构造的特点，通常针叶树材花纹比较简单，阔叶树材花纹则丰富多彩。

3.2　木材的主要显微构造

从植物进化来看，裸子植物早于被子植物，因此针叶树材保留了相对原始的细胞结构，而阔叶树材的解剖结构更为特殊和复杂。

3.2.1　针叶树材的显微构造

针叶树材的解剖分子相对简单，排列较为规整，主要有轴向管胞、木射线、轴向薄壁组织和树脂道。

轴向管胞（longitudinal tracheid 或 tracheid）是指针叶树材中轴向排列的厚壁细胞，同时起着输导水分和机械支撑的作用。轴向管胞可以想象成一根两端捏紧的苏打水吸管。以此类比，在外形和尺寸比例上，吸管和轴向管胞很相似。晚材管胞比早材管胞长且厚。在早材管胞径切面上，纹孔大而多；在晚材管胞径切面上，纹孔小而少。相邻管胞之间通过纹孔相连，形成物质交换的弦向通道（图 3.11）。

木射线（wood ray）大部分是由射线薄壁细胞构成，在边材部分，活的薄壁细胞发挥着贮藏营养物质、径向输送水分和营养物质的作用；在心材部分，薄壁细胞已经死亡。木射线可以分为单列木射线和纺锤型木射线（含横向树脂道）。有的树种的射线也具有厚壁细胞，这种构成木射线的厚壁细胞，称为射线管胞（ray tracheid），如松科的松、云杉、落叶松、雪松、铁杉、黄杉等。

轴向薄壁组织（longitudinal parenchyma）是由许多轴向薄壁细胞聚集而成，在立木中主要起着贮藏养分的作用。针叶树材的轴向薄壁细胞是由砖形或等径形、比较短的和具有单纹孔的细胞所组成。针叶树材中的薄壁组织含量甚少或无，占木材总体积不足 1.5%，仅在罗汉松科、杉科、柏科中含量较多，为该类木材的重要特征。

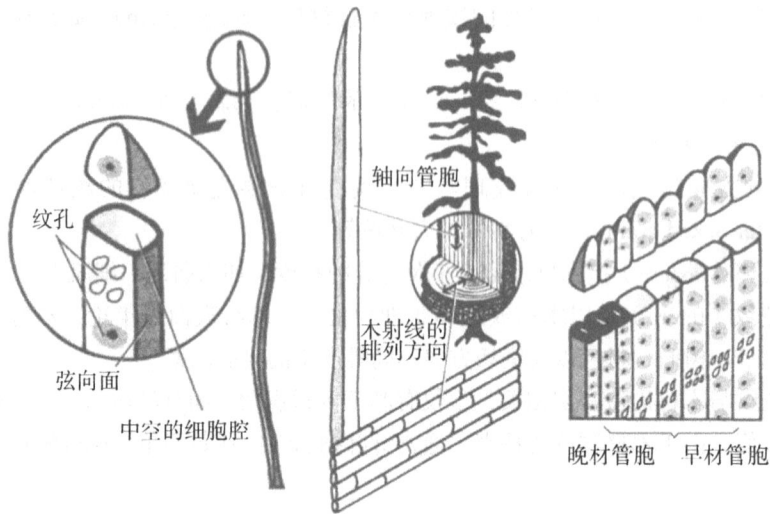

图 3.11　针叶树立木中的细胞排列

　　树脂道（resin canal）是针叶树材中具有分泌树脂功能的一种组织，由生活着的薄壁细胞的幼小组织分离而成，为针叶树材重要的构造之一，占木材体积的 0.1%~0.7%。根据树脂道的发生和发展分为正常树脂道（normal resin canal）和创伤树脂道（traumatic resin canal），但并非所有针叶树种都具有正常树脂道，仅在松科松属、落叶松属、云杉属、黄杉属、银杉属、油杉属木材中具有正常树脂道。

　　图 3.12 为马尾松木材三维显微构造。在横切面上，可以看到许多轴向管胞和一个年轮。在左侧的早材管胞腔大壁薄，在右侧的晚材管胞腔小壁厚。树脂道分为轴向树脂道（longitudinal resin canal）和横向树脂道（transverse resin canal）。在横切面上，可以看到轴向树脂道散布于年轮的晚材或晚材附近。木射线沿半径方向呈辐射状穿过年轮，显示出木射线的宽度和长度。在径切面上，可以看到早材管胞径面上的纹孔较多。在木射线和轴向管胞的交界处，木射线呈带状。显示其长度和高度。在弦切面上，可以看到木射线呈现短竖线或纺锤形，纺锤形木射线是因为横向树脂道的存在。

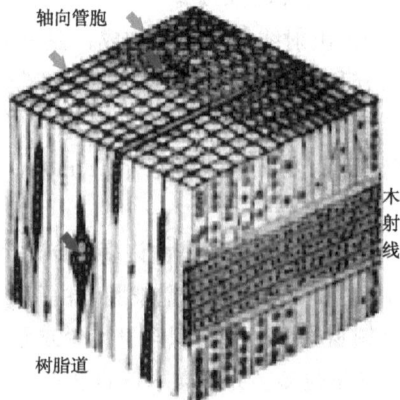

图 3.12　马尾松木材显微构造

3.2.2　阔叶树材的显微构造

阔叶树材除少数树种如水青树、昆兰树外都具有导管（仅限于国产材），因此称为有孔材。组成阔叶树材的分子种类较多，主要有导管（20%）、木纤维（50%）和木射线（17%）等。

阔叶树材组成分子的分类和功能如图 3.13 所示。

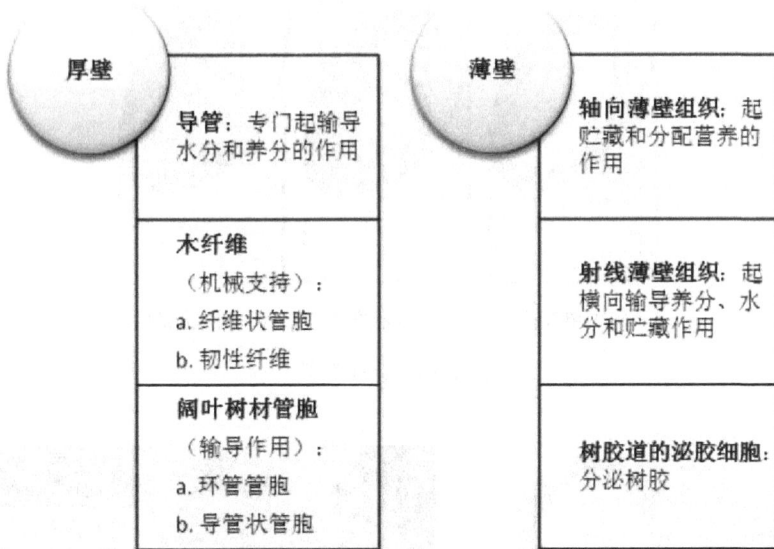

厚壁

导管：专门起输导水分和养分的作用

木纤维

（机械支持）：

a. 纤维状管胞

b. 韧性纤维

阔叶树材管胞

（输导作用）：

a. 环管管胞

b. 导管状管胞

薄壁

轴向薄壁组织：起贮藏和分配营养的作用

射线薄壁组织：起横向输导养分、水分和贮藏作用

树胶道的泌胶细胞：分泌树胶

图 3.13　阔叶树材主要分子及其在立木中的作用

导管[①]是由一连串轴向细胞形成的无一定长度的管状组织。单个细胞成为导管分子（vessel element）。两个导管分子之间底壁相通的孔隙称为**穿孔**（perforation），底壁相连部分的细胞壁称为穿孔板（perforation plate），其按穿孔形状可以分为：单穿孔（simple perforation）、梯状穿孔（scalariform perforation）、网状穿孔（foraminate perforation）（图 3.14）。

导管的内含物主要为侵填体（tyloses）与树胶（gum），以侵填体为常见。**侵填体**只能形成于导管与薄壁组织相邻之处，它是薄壁组织具有生活力时，由导管周围的薄壁细胞或射线薄壁细胞，经过纹孔口而挤入导管内，以至填塞细胞腔的一部分或全部而形成的。侵填体多见于心材。具侵填体的树种，一般耐久性较高。白橡由于侵填体堵塞了导管，所以木材渗透性小，适用于桶材和船舶用材（图 3.15）。

木纤维（fiber）是两端尖锐，形似纺锤形，腔小壁厚的细胞。主要功能是支撑树体，承受机械作用。根据木纤维胞壁上的纹孔类型，木纤维可以分为胞壁有具缘纹孔的**纤维状管胞**（fiber tracheid）和有单纹孔的**韧型纤维**（libriform fiber）。这两种纤维可以存在于同

---

① 导管在木材横切面上呈现孔状，称为管孔。根据管孔分布状态，可将木材分为环孔材、半环孔材和散孔材。管孔相互之间的组合有 4 种：单管孔、复管孔、管孔链和管孔团。

一树种中。有些树种还可能存在一些特殊纤维，如分隔木纤维（septate wood fiber）和胶质木纤维（gelatinous fiber）。分隔木纤维常见于热带木材，如桃花心木。而胶质木纤维是应拉木的特征之一，如杨木。

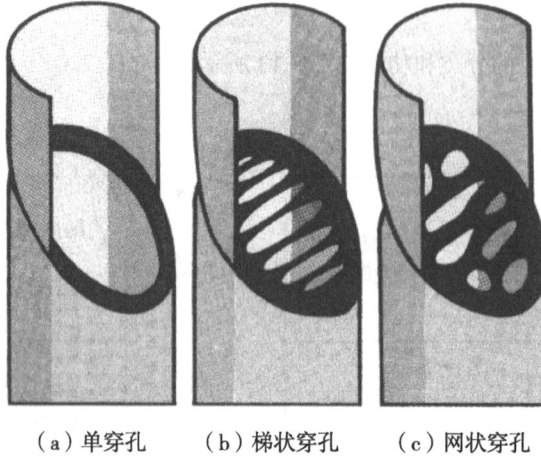

（a）单穿孔　　　（b）梯状穿孔　　　（c）网状穿孔

图3.14　穿孔类型

图3.15　富含侵填体的白橡

轴向薄壁组织由形成层纺锤形原始细胞衍生成2个或2个以上的具单纹孔的薄壁细胞，纵向串联而成的轴向组织（图3.16）。其主要功能是贮藏和分配养分。根据轴向薄壁细胞和导管连生的关系，可以分为离管型（apotracheal parenchyma）和傍管型（paratracheal parenchyma）。

与针叶树材比，阔叶树材的**木射线**相对较宽且相当复杂，是识别阔叶树材的一个重要特征。根据木射线在旋切面上的宽度（射线细胞个数），可以分为单列（uniseriate）、双列（biseriate）、多列（multiseriate）及聚合（aggregate）四种。

树胶道（gum canal）分为轴向和径向，但同时具有轴向和径向两种树胶道者极少，仅限于龙脑香科、苏木科的极少数树种。此外，阔叶树材中也存在正常树胶道（normal gum canal）和创伤树胶道（traumatic gum canal）。

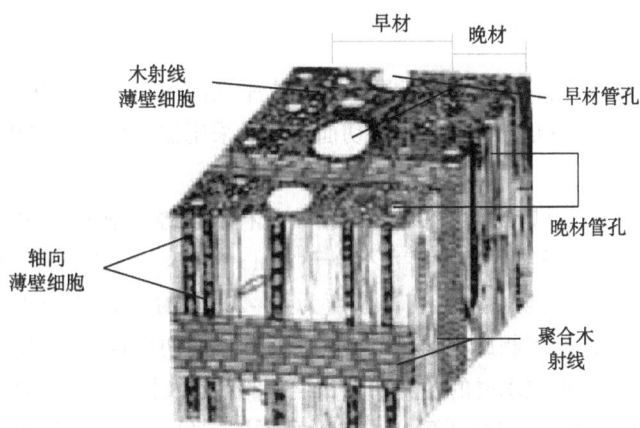

图 3.16　阔叶树材显微构造

3.3　木材识别方法

木材识别是以木材的构造特征为依据，对木材的树种进行识别。传统识别方法主要根据木材的宏观和显微构造特征进行木材树种的识别。

3.3.1　宏观特征识别方法

宏观特征识别主要通过肉眼或者 10 倍放大镜观察木材试样（图 3.17）。一般先用锋利的小刀削光木样部分端面，并用清水润湿，通过观察木材外部表观特征及横切面宏观解剖特征来识别木材。从横切面可观察木材心边材、生长轮及导管、射线与轴向薄壁组织的大小和排列方式，同时结合木材的材色、光泽、纹理、结构、花纹、气味与滋味和木材重量及硬度等特征来综合判断鉴别木材。

宏观识别方法简单易行，在木材加工生产和流通现场及海关和质检等质检现场中只能采用这种方法。但对难以区分的不常见树种或者是部分进口的热带木材，此方法只能识别到类。

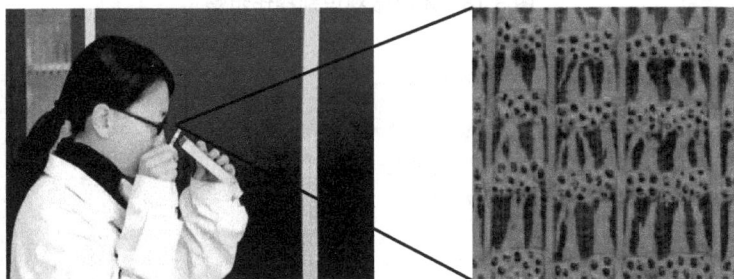

图 3.17　放大镜的正确使用及 10 倍放大镜橡木横切面

3.3.2 微观特征识别方法

由于木材是各向异性的天然材料，要从细胞水平全面认识与了解其结构，首先需要确定木样的三切面，即横切面、径切面和弦切面。然后采用**徒手切片**或**切片机切片**方式进行制片。

◎ **徒手法**：用锋利刀具（美工刀或刀片）在试样表面轻轻拖过，切成小而薄的切片。

◎ **切片法**：按照木样纹理切取尺寸不小于 2cm × 2cm × 2cm 的试样。将试样用水煮或者化学药剂软化后，放在切片机上切出厚度为 10~20μm 的 3 个切面的切片，再经染色、脱水、透明和封片等流程可制成光学显微切片。将制备的切片置于光学显微镜下，观察其木材微观解剖特征，即组成木材的各类细胞和组织的形态及排列特征（图 3.18）。

按照图 3.18 进行操作，可以得到如图 3.19 所示的木材三切面微观解剖结构。

微观识别涉及的识别特征较多，极大地提高识别的准确性，是目前应用最多，也是最成熟的木材识别方法。

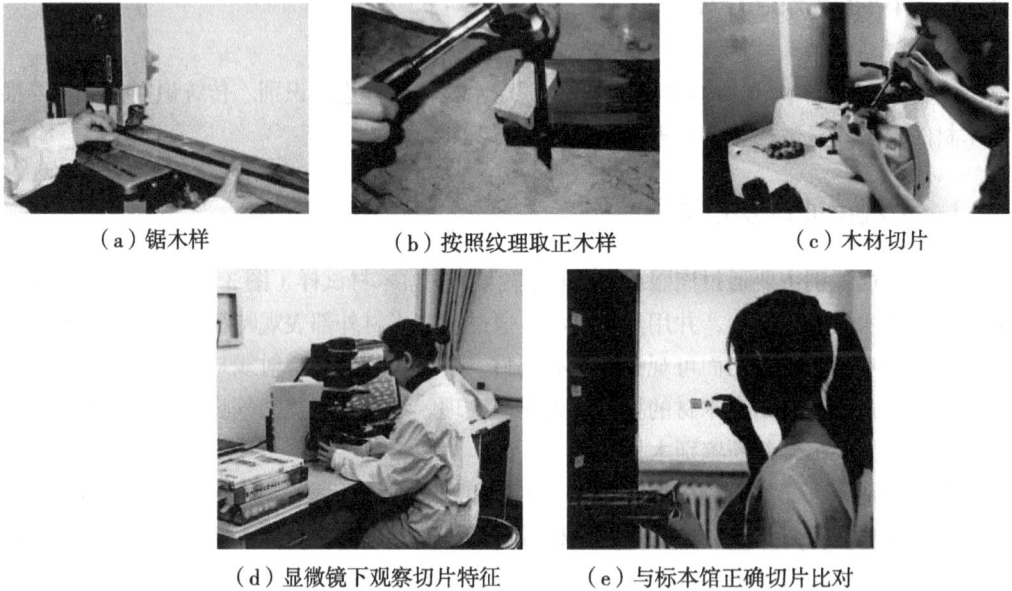

（a）锯木样　　　　　　（b）按照纹理取正木样　　　　　　（c）木材切片

（d）显微镜下观察切片特征　　　（e）与标本馆正确切片比对

图 3.18　木材微观识别操作步骤

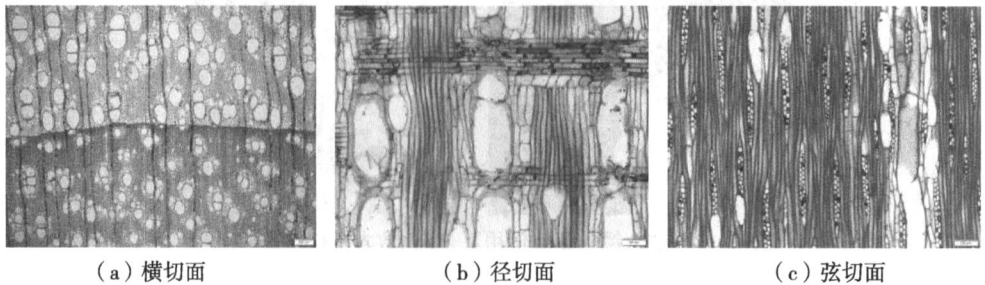

（a）横切面　　　　　　（b）径切面　　　　　　（c）弦切面

图 3.19　显微镜下香樟木三切面微观解剖结构

3.4　主要商品材介绍

3.4.1　常用国产木材

（1）杉木 [*Cunninghamia lanceolata*（Lamb.）Hook.]

杉科杉木属，英文名 Chinese fir，是我国特有的人工速生林树种。木材黄白色，有时心材呈淡红褐色，质较软，细致，有香气，纹理直，易加工，气干密度 0.38g/cm³，耐腐力强，不受白蚁蛀食，被广泛应用于建筑、桥梁、门窗、家具、装修、木制工艺品等（图 3.20）。

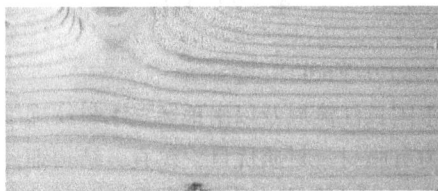

图 3.20　杉木

（2）马尾松（*Pinus massoniana* Lamb.）

松科松属，又称青松、山松。心边材区别不明显，淡黄褐色，纹理直，结构粗，气干密度 0.39~0.49g/cm³，有弹性，富树脂，耐腐力弱。供建筑、枕木、矿柱、家具及木纤维工业（人造丝浆及造纸）原料等用材（图 3.21）。

图 3.21　马尾松

（3）落叶松 [*Larix gmelinii*（Rupr.）Kuzen.]

松科落叶松属，是中国大兴安岭针叶林的主要树种，木材蓄积丰富。木材略重，硬度中等，易裂，边材淡黄色，心材黄褐色至红褐色，纹理直，结构细密，气干密度 0.32~0.52g/cm³，有树脂，耐久用。可供房屋建筑、土木工程、电杆、舟车、细木加工及木纤维工业原料等用材（图 3.22）。

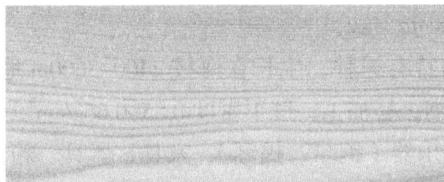

图 3.22　落叶松

（4）红松（*Pinus koraiensis* Sieb. et Zucc.）

松科松属，高可达 40m。小兴安岭被誉为"红松故乡"。木材轻软、细致、纹理直、耐腐蚀性强，为建筑、桥梁、枕木、家具优良用材（图 3.23）。

图 3.23　红松

（5）柏木（*Cupressus funebris* Endl.）

柏科柏木属，分布于中国长江流域及以南地区。木材纹理细，质坚，能耐水，常见于庙宇、殿堂、庭院。木材为有脂材，材质优良，纹直，结构细，耐腐，是建筑、车船、桥梁、家具和器具等用材（图 3.24），最早应用于汉广陵王墓中"黄肠题凑"。

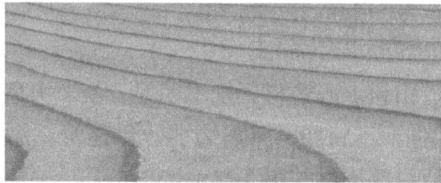

图 3.24　柏木

（6）樟子松（*Pinus sylvestris* var. *mongolica* Litv.）

松科松属，产于中国黑龙江大兴安岭海拔 400~900m 山地及海拉尔以西、以南一带沙丘地区。樟子松是东北地区主要速生用材、防护绿化、水土保持优良树种。材质较强，纹理直，可供建筑、家具等用材（图 3.25）。

图 3.25　樟子松

（7）云杉（*Picea asperata* Mast.）

松科云杉属，为中国特有树种，生长在海拔 2400~3600m 地带。木材节少，黄白色，材质轻软，纹理直，结构细，易加工，气干密度 0.55~0.66g/cm³，有弹性，具有良好的共鸣性能。可供建筑、飞机、乐器（钢琴、提琴）、家具、器具、箱盒、刨制胶合板与薄木以及木纤维工业原料等用材（图 3.26）。

图 3.26 云杉

（8）香樟木 [*Cinnamomum camphora*（L.）Presl.]

樟科樟属，主要分布于中国长江流域以南区域，是我国亚热带常绿阔叶林的重要树种。香樟木作为中国传统的名贵树种木材，具有纹理美观、木质坚硬，散发浓郁芳香、耐腐朽和防虫蛀等优点（图3.27），与梓木、楠木和稠木并称"江南四大名木"。香樟木是制作高级建筑、造船、高级家具和雕刻的理想良材。

图 3.27 香樟木

（9）水曲柳（*Fraxinus mandshurica* Rupr.）

木犀科梣属，分布于中国东北、华北等地区和陕西、甘肃、湖北等省。生于海拔700~2100m 的山坡疏林中或河谷平缓山地。朝鲜、俄罗斯、日本也有分布。水曲柳第三纪孑遗种。与胡桃楸、黄波罗被称为中国东北珍贵的"三大硬阔树种"，其木材坚硬致密，纹理美观，是工业和民用的高级用材（图3.28）。

图 3.28 水曲柳

（10）柞木 [*Xylosma racemosum*（Sieb. et Zucc.）Miq.]

大风子科柞木属，国家二级珍贵树种，主产秦岭以南和长江以南各省区。材质坚硬，气干密度 0.63~0.72g/cm^3，纹理美观，具有抗腐耐水湿等特点（图3.29）。

图 3.29 柞木

（11）胡桃楸（*Juglans mandshurica* Maxim.）

胡桃科胡桃属，又称核桃楸。胡桃楸材质坚硬、致密，纹理通直、耐腐、弹性好、加工容易、刨面光滑、油漆性能好，广泛应用于建筑、船舰、车辆装修、军工、家具等方面（图3.30）。

图3.30　胡桃楸

（12）黄波罗（*Phellodendon amurense* Rupr.）

芸香科黄檗属，黄波罗木材性质坚硬，耐水湿，耐腐力强，纹理美丽，有光泽，材质软，易加工（图3.31）。可供枪托、飞机、造船、建筑、胶合板及家具所用。

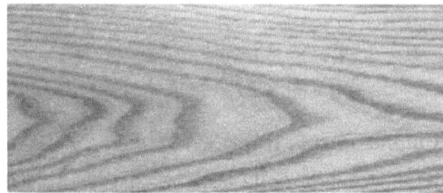

图3.31　黄波罗

（13）青冈 [*Quercus glauca*（Thunb.）Oerst]

壳斗科青冈属，又称青冈栎、紫心木。该种木材坚韧，可供桩柱、车船、工具柄等用材（图3.32）。

图3.32　青冈

（14）速生材

速生材树种是指生长快、成材早、轮伐期短的树木品种，我国常见的阔叶树有刺槐、泡桐、杨树、桉树、柳树和针叶树的柳杉等。由这些速生树种制成的木材称为速生材，通常质地较软，一般用作造纸、建筑、人造板原料等，也有部分用于制造家具，但都必须要对木材作一些处理。

3.4.2 常用进口木材

（1）橡木（*Quercus* spp.）

壳斗科栎属木材，又称栎木。按照国家标准《中国主要进口木材名称》（GB/T
18513—2001），该属木材分**白橡**（*Quercus rubra*）和**红橡**（*Quercus alba*）两类商品材。
橡木为环孔材，构造特征类似于中国产的麻栎和槲栎。

用于家具、地板等室内装饰材料使用时，白橡和红橡均可。但用于密封的桶状制品使
用时，只能使用白橡，不能使用红橡（因为白橡心材管孔内含丰富的侵填体，抗渗性比红
橡好），白橡木材也常用于制作酒桶，用来酿造和保存葡萄酒（图 3.33）。

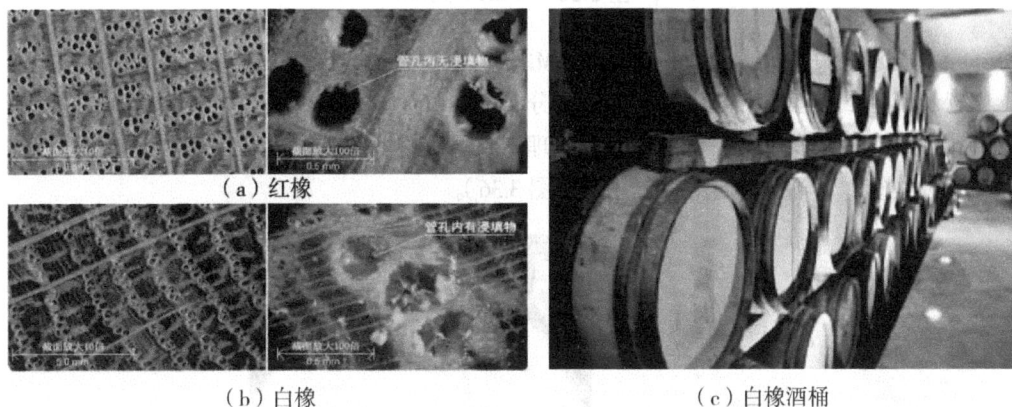

（a）红橡

（b）白橡　　　　　　　　　　（c）白橡酒桶

图 3.33　红橡和白橡的差异及白橡酒桶

（2）山毛榉（*Fagus sylvatica*）

山毛榉，别名水青冈，英文名 Fagus，壳斗科山毛榉属的一种植物，广泛分布在亚洲、
欧洲与北美洲，也是温带阔叶落叶林的主要构成树种之一。

木材心材边材区别不明显。木材呈浅红褐至红褐色，有光泽，无特殊气味（图 3.34）。
木材纹理直或斜，结构中且均匀，材质重且硬。木材气干密度平均值 0.792g/cm³。山毛榉
木材是优良的家具、地板用材。

图 3.34　山毛榉

（3）枫木（*Acer* spp.）

槭树科槭属树种，主要产亚洲、欧洲、北美洲三洲的北温带地区。枫树是加拿大的国

树。木材呈灰褐至灰红色，年轮不明显，管孔多而小，分布均匀（图3.35）。枫木可大致分为2类，即**硬枫**（如糖枫）和**软枫**（如红枫），它是优良的家具、地板用材。

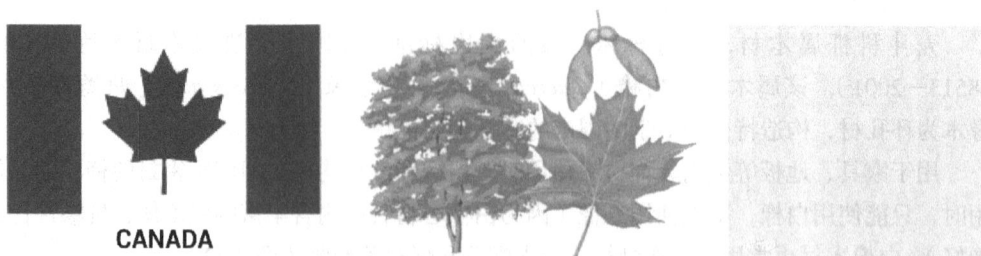

图3.35　加拿大与枫木

（4）花旗松 [*Pseudotsuga menziesii*（Mirb.）Franco]

松科黄杉属，英文名 Douglas fir，原分布于加拿大、墨西哥和美国的太平洋沿岸。中国庐山、北京等地有栽培。材质坚韧，纹理细致，心材淡红色，边材淡黄色，有弹性，耐久用，为建筑、车辆、家具等良好用材（图3.36）。

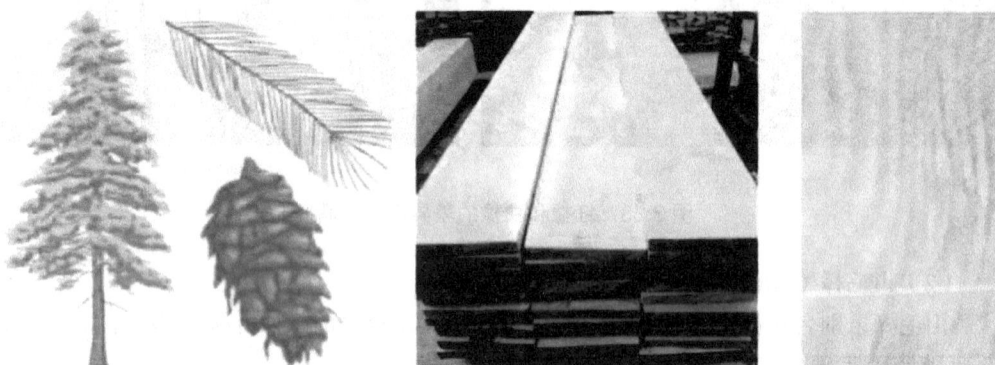

图3.36　花旗松

（5）美国南方松

美国南方松又称南方黄松，是长叶松（大王松 Longleaf pine, *Pinus palustris*）（图3.37）、短叶松（芒刺松 Shortleaf pine, *Pinus banksiana* Lamb.）、湿地松（Slash pines, *Pinus ellitti*）和火炬松（德华松 Loblolly pine, *Pinus taeda*）4个树种集群名称，生长于美国南部广大地区。美国南方松不仅拥有美观漂亮的自然纹理外表，而且在所有软木树种中具有最强韧材质，不易受碰撞损伤并极为耐磨，而且有良好的握钉能力。美国南方松同时也是世界上最适合作防腐浸泡处理的木材，这是因为其自然多孔的细胞结构特征，可以使防腐剂均匀地渗入木材内层，并持久地保存，以抵御各种霉菌、白蚁和其他微生物的侵蚀。

（6）新西兰辐射松（*Pinus radiata* D. Don）

新西兰辐射松，英文名 Radiata pine。辐射松木材为中密度、结构均匀、收缩效率平均、稳定性强的优质软材（图3.38）。完好的原木不存在腐朽、心腐和虫咬等问题；木材握钉力好，渗透性强，极易防腐，干燥、固化和上色等处理。辐射松用途广泛，可以用于建筑、胶合板、纤维板、刨花板、纸浆、家具、电杆、围栏和枕木等领域。

图 3.37　长叶松

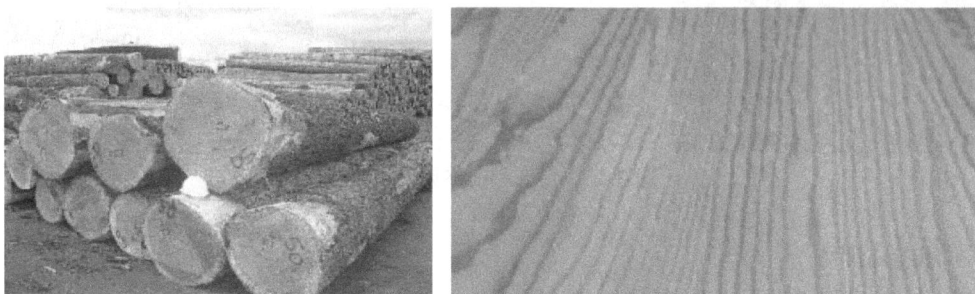

图 3.38　新西兰辐射松

（7）加拿大铁杉 [*Tsuga canadensis*（L.）Carrière]

松科铁杉属，是加拿大卑诗省蕴藏量最丰富的沿海树种。年轮清晰，木色淡雅均匀，纹理笔直均匀，木质细密，硬度适中，具有较高的抗弯强度及自然的防腐能力（图 3.39）。气干密度可达 $0.47g/cm^3$，弹性模量 11.24GPa，抗弯强度 77.9MPa，抗压强度 37.3MPa。目前，用铁杉代替榧木制作高级围棋盘，俗称"新榧"。

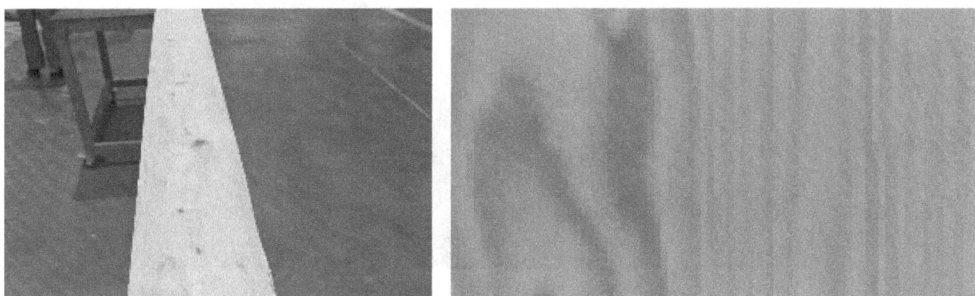

图 3.39　加拿大铁杉

（8）北美 SPF

白云杉（Black spruce）、杰克松（Jack pine）和冷杉（Balsam fir），这 3 个针叶树种是构成 SPF（云杉—松—冷杉）树种组合的主要针叶树种，它们具有许多共同的性状和相同的产地。窑干的 SPF 锯材主要用于各种类型的民用、商用、工业用、农业用建筑的结构框架，同时也广泛用于制造预制房屋、木桁架屋顶和其他结构件，也可用于露台、屋面板、装饰件等。除了用于生产高质量的结构材外，经过少许加工，SPF 木材也可以用来制造非常经济实用的实木家具（图 3.40）。

图 3.40　白云杉、杰克松和冷杉

▶ **思考与训练**

1. 木材横切面上无法看到（　　）特征？

A. 早晚材过渡情况　　B. 木射线的高度　　C. 心材的颜色　　D. 生长轮的宽度

2. 在木质部中，靠近树皮颜色较浅的外环部分称为（　　）。

A. 早材　　　　　　B. 心材　　　　　　C. 晚材　　　　　　D. 边材

3. 仔细查看身边实木家具或器物，并拍一张木材断面照片，判断该材料是针叶树材还是阔叶树材。例如，该直尺断面可以看到管孔，判断用材为阔叶树材，如图所示。

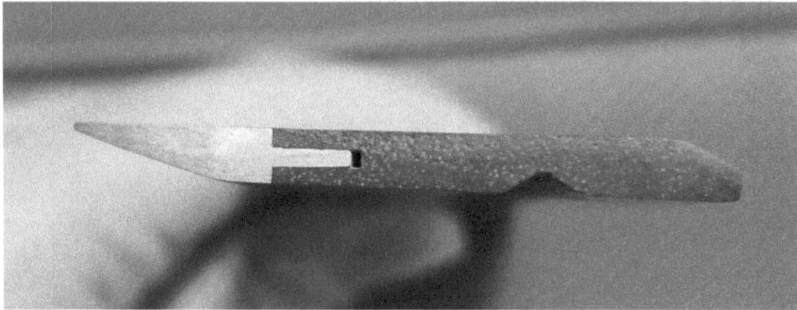

4. 针叶树材和阔叶树材的宏观构造有何不同？

5. 针叶树材和阔叶树材的微观构造有何不同？

6. 分别列举三种国产商品材和三种进口商品材的个性特征。

（1）了解木材细胞壁化学成分；

（2）了解木材细胞壁结构特征；

（3）了解木材的化学利用。

　　木材由碳、氢、氧等化学元素组成，这些元素各自占多大比例？这些元素形成了什么样的化合物？这些化合物在木材体内呈现什么状态？木材纤维素有什么特点？半纤维素和木质素有什么特点？木材抽提物有什么特点？这些主要化学成分如何组成细胞壁结构？它们发挥了什么作用？如何利用这些化学成分？本章将解答这些疑问。

　　从构造上看，木材是由许多中空细长细胞组成的。从化学组成上看，木材细胞壁是一种天然生长的有机高分子复合材料，主要由纤维素、半纤维素、木质素和木材抽提物组成。木材的特性取决于木材化学组成及其细胞壁构造。了解木材的化学组成和细胞壁构造特征有助于认识木材和利用木材。

4.1　木材细胞壁主要成分

　　木材由碳、氢、氧 3 种基本化学元素（chemical element）组成。表 4.1 详细列出了木材的化学元素[①]及各自所占绝干重量的百分比。碳元素的含量大概为木材绝干重量的一半左右，占据主要地位。此外，木材含有无机化合物，这些化合物在高温充分燃烧后依然存在，将这些残留物称为**灰分**（ash）。灰分是含有钙、钾、镁、锰、硅等元素的不燃化合物。从利用的角度来看，国产木材的灰分含量非常低，特别是二氧化硅含量很低，这一点很重要。因为二氧化硅含量超过 0.3%（以干重计算）的木材会使刀具变钝。二氧化硅含量超过 0.5% 的木材常见于热带阔叶树材，某些树种的二氧化硅含量甚至会超过 2%。

---

① 化学元素就是具有相同核电荷数（核内质子数）的一类原子的总称。

表 4.1　木材的元素组成

元素	占比	元素	占比
C	49%	N	0.1%
H	6%	灰分*	0.2%~0.5%
O	44%		

* 在某些热带阔叶树种中灰分高达 3.0%~3.5%。

　　根据其在木材中的含量和作用，把木材的化学成分分为**主要成分**和**次要成分**。**木材的主要化学成分为纤维素、半纤维素和木质素，它们是构成木材细胞壁的主要物质**；次要成分为抽提物和灰分，这两种物质以内含物形式存在于细胞腔中，也有少量存在于细胞壁中，如图 4.1 所示。

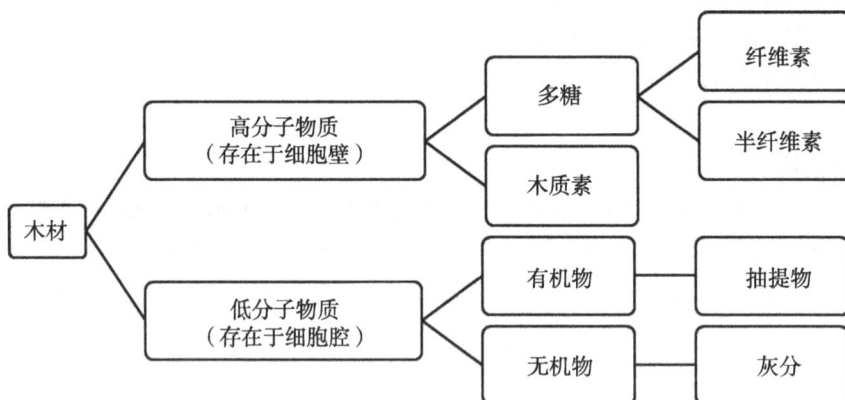

图 4.1　木材的分子组成

　　纤维素、半纤维素和木质素作为细胞壁的主要组成成分，在针叶树材和阔叶树材中的含量不同，见表 4.2。

表 4.2　三大素在针叶树材和阔叶树材中的含量对比

树种	纤维素	半纤维素	木质素
针叶树材（softwood）	（42±2）%	（27±2）%	（28±3）%
阔叶树材（hardwood）	（45±2）%	（30±5）%	（20±4）%

4.1.1　纤维素

（1）纤维素的化学结构

　　纤维素是线性长链聚合物，其化学分子式为（$C_6H_{10}O_5$）$_n$，其聚合度 n 可高达 10 000。葡萄糖和纤维素之间的结构关系如图 4.2 所示。纤维素是由 β-D- **吡喃型葡萄糖**

（glucopyranose）彼此以 β-1-4 苷键、以 C_1 椅式构象联结而成的线形高分子化合物。2 个葡萄糖缩聚形成纤维素二糖（cellubiose）。纤维素的 C、H、O 元素的占比分别为：44.44%、6.17%、49.39%。

纤维素分子链由高达 10 000 个葡萄糖单元组成，貌似结构很大，但其链长只有 10μm（1/1000cm），直径只有 8Å（1Å=1/100 000 000cm），这个尺度依然只能借助电子显微镜才能看到。

根据大量研究，证明了纤维素的化学结构（图 4.2）具有如下特点：

◎ 纤维素大分子仅由一种糖基即葡萄糖基组成，糖基之间以 1-4 苷键连接，即在相邻的两个葡萄糖单元 C_1 和 C_4 之间连接，在酸或高温作用下，苷键会发生断裂，从而使纤维素大分子降解；

◎ 纤维素链的重复单元是纤维素二糖基，其长度为 1.03nm，每一个葡萄糖基与相邻的葡萄糖基之间旋转 180°；

◎ 除两端的葡萄糖基外，中间的每个葡萄糖基上都有 3 个游离羟基，分别位于 C_2、C_3、C_6 位上，它们的反应能力不同，对纤维素的性质具有重要影响；

◎ 纤维素大分子两端的葡萄糖末端基，其结构和性质不同，纤维素分子具有极性和方向性；

◎ 纤维素为结构均匀的线性高分子，除了具有还原性的末端基在一定的条件下氧环式和开链式结构能够互相转换外，其余每个葡萄糖基均为氧环式结构，具有较高的稳定性。

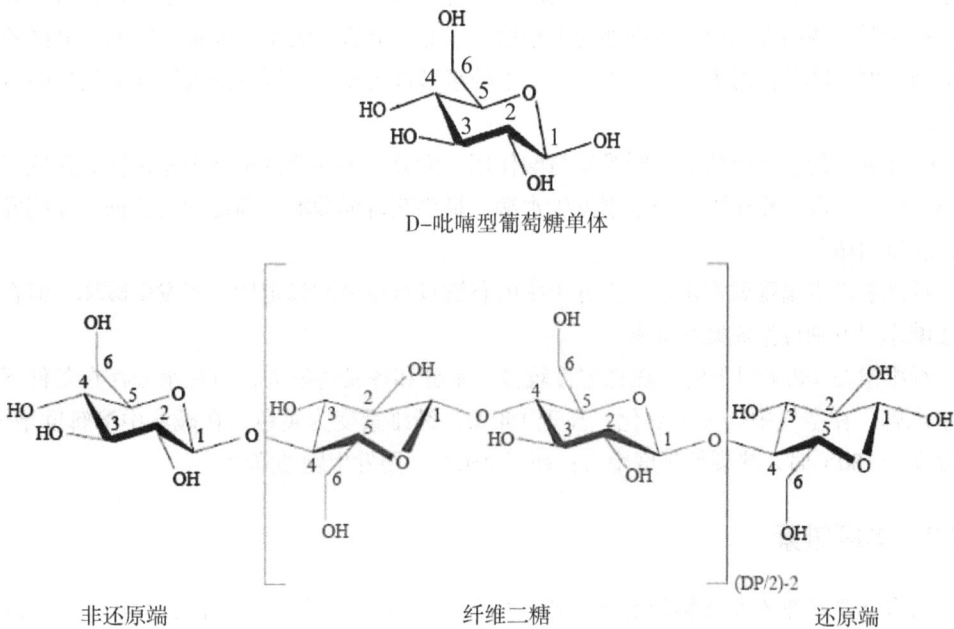

D-吡喃型葡萄糖单体

图 4.2　纤维素分子结构式

（2）纤维素的物理结构

根据X-射线研究，纤维素大分子的聚集体为两相结构，即**纤维素由结晶区（crystalline regions）和非结晶区或无定形区（amorphous regions）交替排列而成**（图4.3）。结晶区分子排列规则、紧密，主要以氢键结合；非结晶区分子排列松散，规则性差；结晶区和非结晶区之间没有明显界限，而是逐步过渡的；非结晶区内的纤维素分子只有一部分形成氢键，另一部分处于游离状态，因而纤维素吸湿发生在非结晶区。

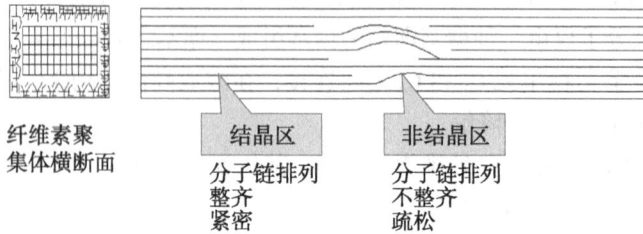

图 4.3　木材纤维素的两相体系

结晶度是指结晶区所占纤维整体的百分率。结晶度大，则木材的抗拉强度、抗弯强度、尺寸的稳定性高。反之，结晶度低，上述性质必然降低，而且木材的吸湿性和化学反应性也随之增强。

（3）纤维素的主要性质

①纤维素的吸湿性　纤维素非结晶区分子链上的羟基，部分形成氢键，部分处于游离状态。游离的羟基为极性基团，易于吸附极性的水分子，与其形成氢键结合，这就是纤维素具有吸湿性的内在原因。吸湿性的大小取决于非结晶区的大小，吸湿性随非结晶区的增加即结晶度的降低而增大。当纤维素分子上的羟基被置换后，纤维的吸湿性则发生明显的变化。

②纤维素的水解或热解　纤维素与酸作用，大分子中的苷键呈现不稳定性。在适当的氢离子浓度、温度作用下，纤维素发生水解，最终获得葡萄糖。葡萄糖为己糖，经酶发酵可以获得酒精。

纤维素在常温稀碱作用下，大分子中的苷键具有较高的稳定性，不发生破坏。但在常温浓碱作用下则可生成碱纤维素。

纤维素在140℃以下时，热稳定性较佳，水分和挥发物散失，但在水分存在条件下会发生水解，在空气中会发生氧化；高于140℃，纤维素变为黄色，在碱液中溶解度增大；温度高于180℃时，热裂解程度增大；超过250℃，则发生剧烈降解。

4.1.2　半纤维素

大多数半纤维素是**支链聚合物**，通常由数百个糖单元组成（聚合度为数百）。半纤维素由不同的几种糖基组成的共聚物，为无定形物质。陆生植物的半纤维素是由 D- 木糖、D- 甘露糖、D- 葡萄糖等组成。

阔叶树材的半纤维素主要是由戊糖组成；针叶树材的半纤维素主要是由己糖组成，己

糖比戊糖稳定。

半纤维素和纤维素同属于多聚糖，同为苷键连接，共存于细胞壁中，具有相近的性质，但也有不同。从其结构来看，区别在于：

◎ 纤维素是单一葡萄糖基构成的均一多聚糖，而半纤维素是由两种或两种以上不同糖基以及少量醛酸基、乙酰基构成的**非均一多聚糖**；

◎ 纤维素是直链型大分子，无支链，而半纤维素主链是线型结构，但具有一个或多个支链；两者的聚合度差异巨大，前者为几千到一万，而后者仅为 150~200，所以说，半纤维素是分子量较低的多聚糖；

◎ 纤维素以微纤丝状态存在细胞壁中，有结晶区和非结晶区之分。一般认为半纤维素不形成微纤丝结构，而且与纤维素之间没有共价键连接，绝大部分存在于非结晶区内与纤维素微纤丝之间通过氢键和范德华力结合。

半纤维素是木材"三大素"中吸湿性最强、耐热性最差、最易分解的组分。

4.1.3 木质素

木质素是具有庞大结构、极其复杂的天然高分子化合物，为无定形物质。木质素不代表单一的物质，而是代表植物中某些性质相同的一类物质。

木质素是具有芳香族特性的，非结晶性的和具有三维空间结构的高聚物，其基本结构单元是苯基丙烷。木质素的 3 种结构单元为：**愈疮木基丙烷、紫丁香基丙烷和对羟苯基丙烷**（图 4.4）。

针叶树材木质素主要是愈疮木基丙烷组成，阔叶树材木质素主要由愈疮木基丙烷和紫丁香基丙烷组成。草本植物（如竹子）含有较多对羟苯基丙烷（表 4.3）。

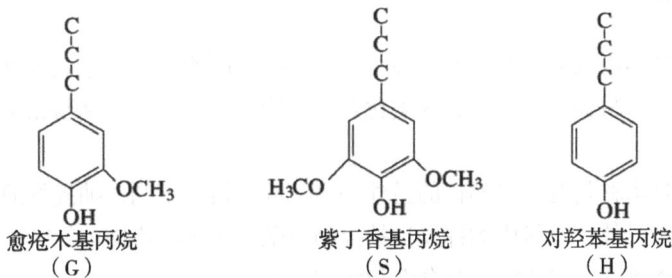

图 4.4　木质素三种基本结构单元

表 4.3　不同材料木质素中 G、S 和 H 的含量

材料	木质素（%）		
	G	S	H
云杉	94	1	5
松	86	2	12
山毛榉	56	40	4
竹材	35	40	25

　　木质素亦可用于鉴别针叶树材和阔叶树材，利用的就是木质素的显色反应（Mäule反应），即将木材试样用1%高锰酸钾溶液处理5min，水洗后用3%盐酸处理，再用水冲洗，然后用浓氨溶液浸透。针叶树材显黄色或黄褐色，阔叶树材则显红色或红紫色。另外，木质素中含有许多发色基团，如苯环、乙烯基等，会影响木材颜色。

　　木质素为无定形聚合物，因而具有**玻璃化转变特性**。当加热木质素达到玻璃化转变温度 T_g 时，木质素迅速软化。当温度高于 T_f，木质素转变为粘流态（图4.5）。木质素的玻璃化转变特性对木质人造板生产、实木弯曲和木材压缩具有重要意义。在高温和水分存在下，木质素易于软化和塑化，发生热软化的温度范围，针叶树材为170~175℃，阔叶树材为160~165℃。

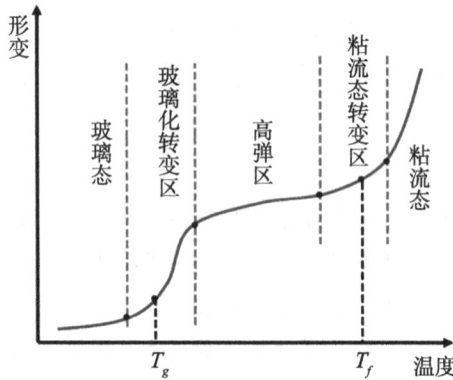

图 4.5　木质素玻璃态转变

4.1.4　木材抽提物

　　木材抽提物是用冷水或热水，及乙醇、苯、乙醚、丙酮或二氯甲烷等有机溶剂抽提出来的物质的总称。木材抽提物主要有3类化合物：脂肪族化合物、萜与萜类化合物、芳香族化合物。

　　木材抽提物是木材具有一定颜色的另外一个原因。树种不同，抽提物的种类和含量不同，则材色不同。另外，木材中的挥发性抽提物也赋予了木材不同的气味。木材抽提物还会影响到木材的渗透性、耐久性、胶合性和涂饰性能。

4.2　木材细胞壁结构

　　木材细胞壁主要由纤维素、半纤维素和木质素构成。如图4.6所示，约40根纤维素大分子以分子链聚集成束形成**基本纤丝**，若干基本纤丝形成**微纤丝**，排列有序的微纤丝状态存在于细胞壁中。

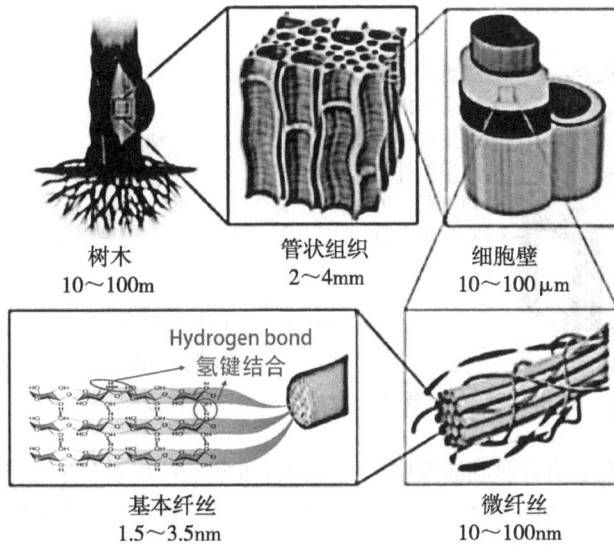

图 4.6 木材细胞壁的多尺度构造

纤维素起着**骨架物质**作用，相当于钢筋水泥构件中的钢筋。半纤维素以无定形状态渗透在骨架物质之中，起着填充作用，故称为**填充物质**，相当于钢筋混凝土中的沙石（图 4.7）。木质素是在细胞分化的最后阶段木质化过程中形成，它渗透在细胞壁的骨架物质和基体物质之中，可使细胞壁坚硬，所以称其为**结壳物质**或**硬固物质**，相当于钢筋混凝土构件中的水泥。由于半纤维和木质素皆为无定形物质，两者共同构成细胞壁的**基体物质**（matrix）。

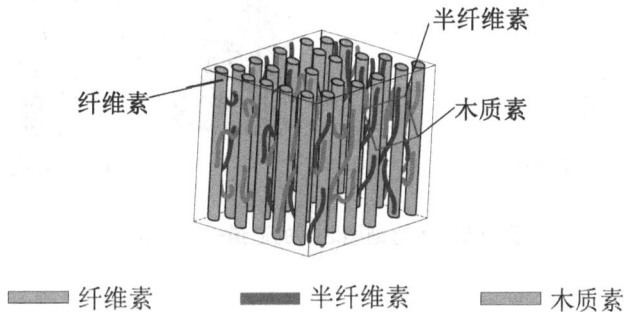

图 4.7 木材细胞壁三大成分的存在状态与作用

在光学显微镜下，根据化学组成的不同，通常将木材细胞壁结构分为**胞间层**（middle lamella，ML）、**初生壁**（primary wall，P）和**次生壁**（secondary wall，S）。其中，次生壁可以分为外层（outer layer，S_1）、中层（middle layer，S_2）及内层（inner layer，S_3）（图 4.8）。

胞间层 ML：是两个相邻细胞中间的一层，为两个细胞共有，很薄，主要由木质素和果胶组成；

初生壁 P：是细胞增大期间形成的壁层，厚度一般为细胞壁厚度的 1% 左右，主要由木质素、纤维素、半纤维素组成；

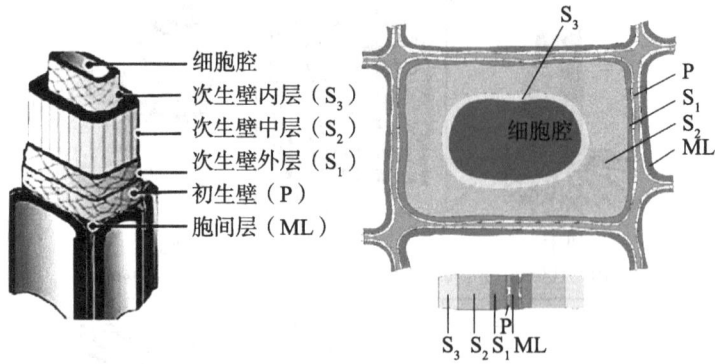

图 4.8　管胞和木纤维细胞壁结构模型

次生壁 S：细胞成熟后形成的壁层，主要由纤维素或纤维素和半纤维素的混合物组成，后期常含有大量木质素和其他物质。由于纤维素组成的微纤丝排列方式不同，次生壁通常分为三层。其中，S_1 层厚度为细胞壁厚度的 10%~22%，微纤丝角（相对于细胞轴向）为 50°~70°；S_2 层厚度为细胞壁厚度的 70%~90%，微纤丝角为 10°~30°，甚至几乎平行；S_3 层厚度为细胞壁厚度的 2%~8%，微纤丝角为 60°~90°。所以，细胞壁次生壁中层（S_2）决定了细胞壁的物理力学性质，进而决定了木材的物理力学性质。

纹孔（pit）通常是指细胞壁的次生壁上凹陷的部分，它是立木中相邻细胞间水分和养料运输的通道；纹孔的渗透性关系到木材干燥、防腐、防火的浸注以及制浆等加工工艺。根据孔腔形式的不同，纹孔可分为单纹孔和具缘纹孔（图 4.9）。

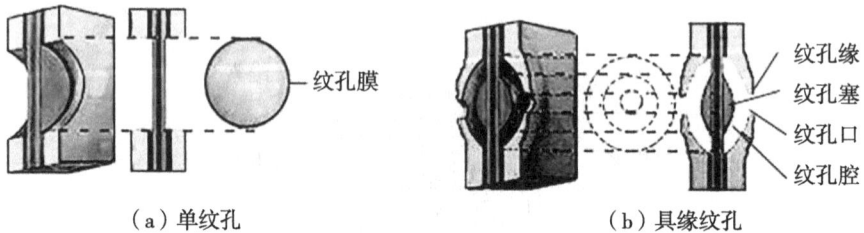

（a）单纹孔　　　　　　　　　　　（b）具缘纹孔

图 4.9　木材细胞之间的通道——纹孔

4.3　木材化学利用举例

4.3.1　制浆造纸

绝大多数纸张是用富含纤维素的木材制备的。木质素在植物中的作用就像胶水，它能把纤维紧紧地粘在一起。造纸需要把原料制成纸浆，加碱蒸煮，主要目的就是要尽量去除木质素，因为残留的木质素会使纸的强度降低。

在造纸过程的制浆工序中，虽然去除了绝大部分木质素，但是不可能把木质素完全除尽，因此还需要漂白，来进一步去除木质素。漂白工艺分为氧化漂白和还原漂白两种。在氧化条件下，残余的木质素转化为可溶性物质，然后被洗脱，所以这一类纸张不易泛黄。而在还原条件下，并不能去除残余的木质素，只是选择性地破坏了有色物质的发色基团，使纸张变白，所以这一类纸张耐久性就比较差，时间长了，受光照和空气氧化，容易泛黄。新闻纸比普通的纸更容易泛黄，主要就是因为没有经过漂白工序，里面所含的木质素含量更高。

4.3.2　木材焊接

木材摩擦焊接是指木材在外力的作用下，通过摩擦生热，使半纤维素和木质素受热软化、融合，在交界层形成交联网状结构。根据焊接方式的不同，木材摩擦焊接又分为线性摩擦焊接和旋转摩擦焊接（图 4.10）。线性摩擦焊接是指在木材表面施加压力，2 块木质基材在一定振幅和频率下进行往复地线性摩擦运动，相邻面的木质素和半纤维素融合，并在保压冷却后形成连接强度的焊接方式。旋转摩擦焊接是指在一定压力下将旋转的圆木榫插入预留钻孔，在圆木榫外侧与钻孔内侧形成熔融界面，进一步冷却后实现连接的焊接方式。

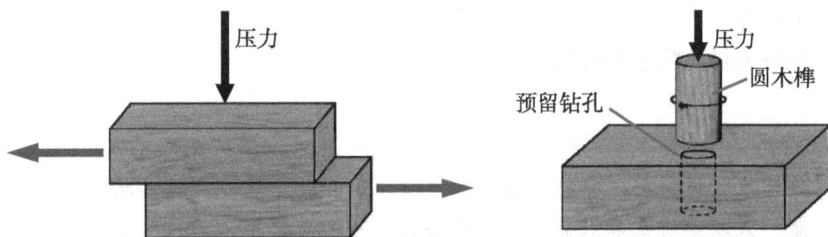

图 4.10　木材线性摩擦焊接和旋转摩擦焊接

> 思考与训练

1. 木材细胞壁结构中，最厚的是哪一部分？
2. 木材细胞壁的化学成分中起骨架作用的是（　　）。
　A. 纤维素　　B. 半纤维素　　C. 木质素　　D. 果胶
3. 纤维素、半纤维素、木质素在细胞壁结构中的作用是什么？
4. 纤维素的结构单元是什么？它有什么特点？
5. 决定木材吸湿性的化学成分是什么？
6. 木质素的结构单元是什么？
7. 请列举木材三大素的利用（不限于本教材介绍）。
8. 借助软件或手工绘制细胞壁的壁层结构，指明各部分名称。
9. 为什么说细胞壁次生壁中层（S_2）决定了细胞壁和木材的物理力学性质？

第**5**章
木材物理性质

▶ **学习目标**

（1）了解木材中水分的存在状态和含水率测定方法；

（2）理解纤维饱和点的概念；

（3）理解木材干缩湿胀特性；

（4）了解提高木材尺寸稳定性的措施；

（5）理解木材孔隙率与密度。

▶ **本章描述**

　　木材作为生物质材料，具有独特的构造特征和生物学特征，这些特征赋予了木材特殊的物理性质。木材中有哪些水分，这些水分存在于哪里？含水率这一关键物理指标如何测定？木材干缩湿胀特性指的是什么？在应用中会带来哪些问题？木材密度如何界定？本任务将解答这些疑问。

　　木材的物理性质是指既不改变木材化学成分，又不破坏木材的完整性而表现出来的性质，它包括密度、干缩和木材与水、热、电等物理现象发生关系时表现出来的性能。木材的物理性质对木材的利用有至关重要的影响。

5.1　木材水分相关性质

5.1.1　木材中的水分

　　树木生长从根部吸收水分，通过边材输送到树木各部，树叶光合作用所制造的养分由韧皮部输送到树木各部分，每生长 $1m^3$ 木材，大约要从土壤中蒸腾出 $300{\sim}400m^3$ 的水分。立木中水分是树木生长中必不可少的物质，又是树干输送各种物质的载体。此外，由于木材是多孔性材料，在水存、水运、水热处理过程中，水均可渗入木材内部，干木材还能从空气中吸收蒸汽状态的水分。

按京都大学中户莞的分类方法，将木材孔隙分为两类。

◎ 永久孔隙：干燥、湿润状态或被其他润胀剂充满时，其孔大小几乎不发生变化。如细胞腔。

◎ 瞬时孔隙：由润胀剂暂时生成，干燥时消失的孔隙。如细胞壁中孔隙，水分吸着时形成，解吸时消失者。

木材中的水分按其存在状态可分为**自由水**（free water）和**吸着水**（bound water）（图 5.1）。

◎ 自由水：是指游离态存在于木材永久孔隙中的水分。自由水的含量主要由木材孔隙度决定，它影响到木材重量[①]、燃烧性、渗透性和耐久性，对木材体积稳定性、力学、电学等性质基本上无太大影响。

◎ 吸着水：是指以吸附状态存在于木材瞬间孔隙中的水，即细胞壁微纤丝之间的水分。吸着水多少对木材物理力学性质和木材加工利用有重要的影响。

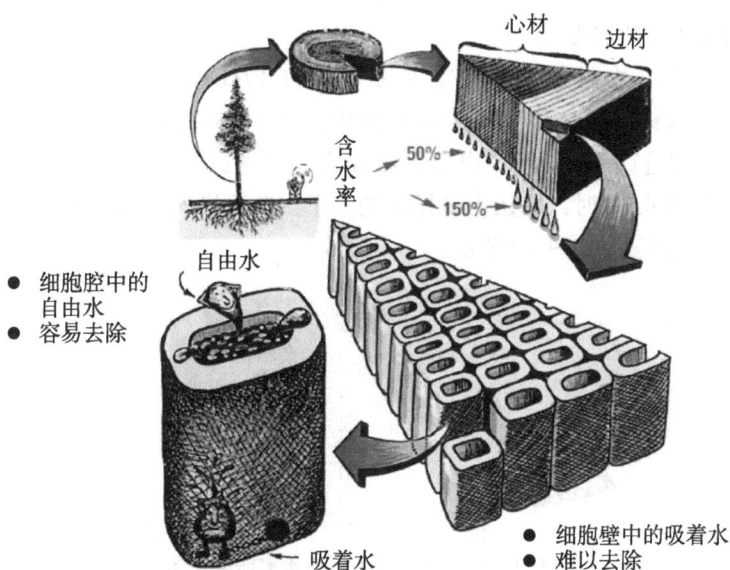

图 5.1　木材中水分的存在状态

细胞壁内的吸着水由吸附力（主要是氢键）控制。当谈到合成海绵**吸收**（absorption）水分时，不应和"吸附（adsorption）"的概念相混淆。吸收是由**表面张力**（surface tension）和**毛细管力**（capillary force）引起的，它会导致多孔木材中累积大量的水。相反，**吸附**是指水分子被存在于纤维素、半纤维素和木质素中的游离羟基所吸引形成氢键结合。氢键结合发生在整个木材化学成分中羟基（—OH）的氢原子一侧。

---

① 木材重量，此处指的是木材的克重。严格来讲，应该用质量这一物理学概念。但是，为了和衡量木材品质优劣所用的"质量"相区分，本书中统一用木材重量来表示木材的克重，用木材质量表示木材品质的优劣。

5.1.2　木材含水率

通常讲木材的干湿是用来定性描述木材中水分的含量。而要定量描述，则需引入木材含水率的概念。木材中水分的重量和木材自身重量的比值用百分率来表示，称之为木材含水率。

（1）木材含水率的计算

根据计算基准不同，含水率可分为**绝对含水率**（absolute moisture content）和相对含水率（relative moisture content），木材工业中常采用**绝对含水率**。

若以绝干木材的重量 m_0 为计算基准，所得含水率称为绝对含水率，公式为

$$MC = \frac{m_1 - m_0}{m_0} \times 100\%$$

若以湿木材的重量 m_1 为计算基准，所得含水率称为相对含水率，公式为

$$MC = \frac{m_1 - m_0}{m_1} \times 100\%$$

（2）木材含水率的测定

根据国家标准《木材含水率测定方法》（GB/T 1931—2009），制备尺寸为 20mm × 20mm × 20mm 的木材试样，将试样放入烘箱内，在（103±2）℃的温度下烘 8h 后，从中选定 2~3 个试样进行一次试称，以后每隔 2h 称量所选试样一次，至最后两次称量之差不超过试样重量的 0.5% 时，即认为试样达到绝干，然后代入含水率计算公式进行计算。木材含水率测定所用设备如图 5.2 所示。

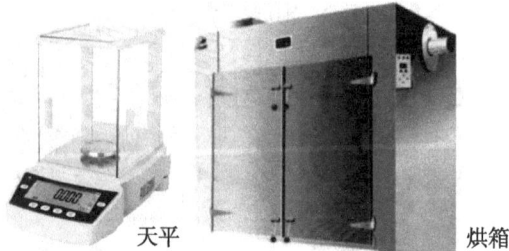

天平　　　　　　　　　烘箱

图 5.2　称重法测定木材含水率所需设备

（3）平衡含水率

木材中水分含量的多少与所处环境的相对湿度和温度有很大关系。当空气中的水蒸气压力大于木材表面水蒸气[1]压力时，木材从空气中吸附水分，这种现象称为**吸湿**（adsorption）；反之，若空气的水蒸气压力小于木材表面的水蒸气压力时，木材向空气中蒸发水分，这种现象称为**解吸**（desorption）。吸湿与解吸仅指吸着水的吸附和排除。

木材长期暴露在一定温度和湿度的空气中，最后会达到相对恒定的含水率，即蒸发水分和吸附水分的速度相等，此时木材所具有的含水率称为**平衡含水率**（equilibrium moisture content，EMC）。

---

[1]　水蒸气，是水（H_2O）的气体形式。当水达到沸点时，水就变成水蒸气。在海平面处一个标准大气压下，水的沸点为 99.974℃。当水在沸点以下时，水也可以缓慢蒸发成水蒸气。而在极低压环境下（小于 0.006 大气压），冰会直接升华变水蒸气。水蒸气指特定空间的水以气、液二态同时存在。

木材的平衡含水率主要取决于大气的温度和湿度，在一定的温度和湿度条件下，平衡含水率就恒定。如温度为 20℃，相对湿度为 35% 时，平衡含水率为 7%；当相对湿度为 70% 时，平衡含水率为 13%；当相对湿度为 65% 时，平衡含水率为 12%。

木材从高湿度侧到达的平衡含水率称为解吸平衡含水率（MC_d）。从低湿度侧到达的平衡含水率称为吸湿平衡含水率（MC_a）。在同一相对湿度下，解吸平衡含水率总大于吸湿平衡含水率（$\Delta MC = MC_d - MC_a > 0$），把这种现象称为**吸湿滞后**现象（sorption hysteresis）（图 5.3）。

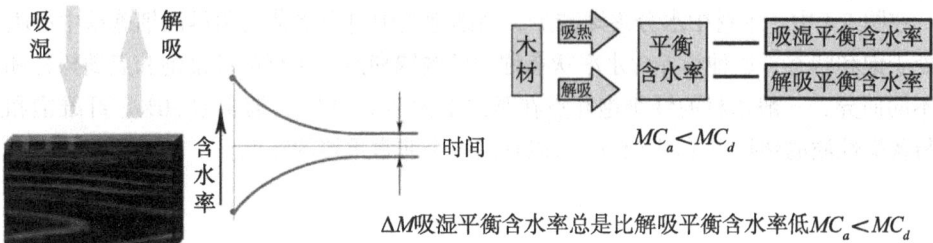

图 5.3　吸湿滞后

木粉、单板等细或薄的木料吸湿滞后数值极小，可忽略不计。对于气干材，吸湿滞后数值不大，一般可粗略认为：$MC_d = MC_a = MC_E$；而窑干材吸湿滞后 ΔMC 大，通常在 1%~5%，取其平均值 2.5%，因此，$MC_E = MC_d = MC_a + 2.5\%$。在木材人工干燥时，应将木材干燥至平衡含水率以下，即 $MC_f = MC_a = MC_E - 2.5\%$。

含水率对木材性能影响很大，我国以往木材物理和力学性质标准中以 15% 为标准含水率，为了与国际接轨，目前以 12% 为标准含水率。

（4）纤维饱和点

随着含水率的降低，木材细胞中水分状态分别经历了饱水状态、生材状态、**纤维饱和点**（fiber saturation point）、气干状态、绝干状态等状态（图 5.4）。不同含水率状态的木材分别称作：湿材、生材、气干材（窑干材）和绝干材。

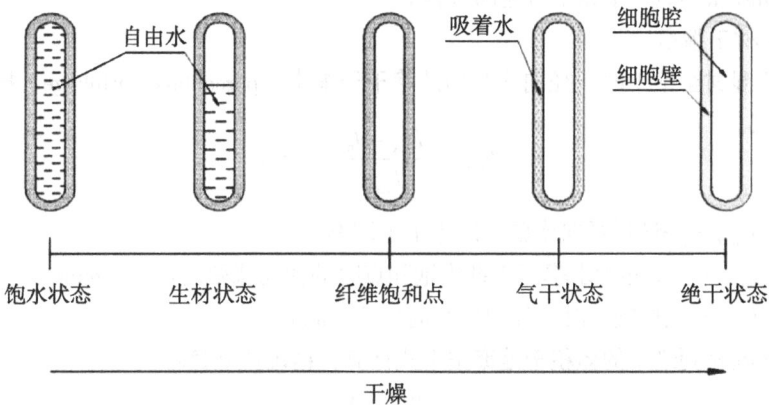

图 5.4　木材细胞各种含水率状态

湿材：长期浸泡在水中的木材称为**湿材**（wet wood），此时细胞处于饱水状态，含水率很高。

生材：树木新伐倒后获得的木材称为**生材**（green wood），含水率在 50% 以上。

气干材：将生材或湿材放置在大气中，水分逐渐蒸发，最后与大气湿度和温度平衡时的木材称为**气干材**（air-dry wood），此时的木材含水率称之为平衡含水率。

窑干材：经过人工干燥的木材称之为**窑干材**（kiln-dry wood）。

绝干材：将木材干燥至含水率为零，这种含水率状态下的木材称之为**绝干材**（oven-dry wood）。

在图 5.4 中，木材中水分逐渐减少，当细胞腔中自由水蒸发殆尽而细胞壁中的吸着水还处于饱和状态，这种临界含水率状态称为**纤维饱和点**。木材的纤维饱和点随树种和温度的不同而异，一般木材的纤维饱和点在常温下为 24%~42%，通常取 30%。纤维饱和点是木材各类性质的转折点（图 5.5），所以它是一个非常重要的特性。

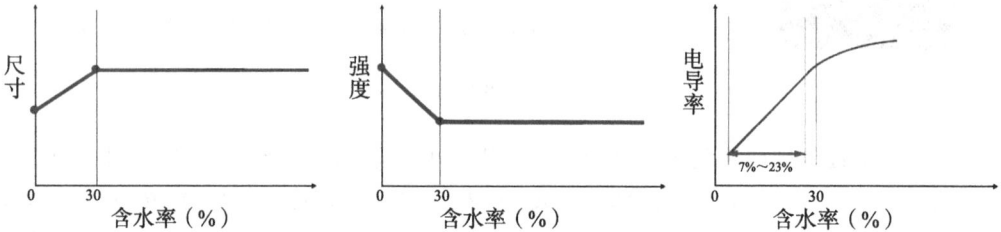

图 5.5　纤维饱和点与木材性质的关系

5.1.3　木材干缩与湿胀

木材含水率低于纤维饱和点时，含水率降低（解吸）会引起它尺寸和体积的缩小，称之为木材**干缩**（shrinkage）。相反，含水率的增加（吸湿）会引起尺寸和体积的膨胀，称之为木材**湿胀**（swelling）。干缩和湿胀并不是在任何含水率条件下都能发生的，而只有在纤维饱和点以下才会发生。在木材的 3 个方向上，干缩和湿胀程度不同。另外要注意的是，干缩和湿胀不是一个完全可逆的过程。

（1）木材干缩率

试样从湿材至绝干时，径向或弦向的**绝干干缩率**（percentage shrinkage）按下式计算，精确至 0.1%。

$$\beta_{max} = \frac{l_{max} - l_0}{l_{max}} \times 100\%$$

式中：β_{max}——试样径向或弦向的绝干干缩率；

l_{max}——试样含水率高于纤维饱和点时的径向或弦向尺寸，mm；

l_0——试样绝干时的径向或弦向尺寸，mm。

试样从湿材到绝干的**体积干缩率**按下式计算，精确至 0.1%。

$$\beta_{v\,max} = \frac{v_{max} - v_0}{v_{max}} \times 100\%$$

式中：β_{vmax}——试样的体积干缩率；

v_{max}——试样含水率高于纤维饱和点时的体积，mm^3；

v_0——试样绝干时的体积，mm^3。

通常，木材纵向的绝干干缩率仅为 0.1%~0.37%，径向为 3%~6%，弦向为 6%~12%。可见，木材横向干缩较纵向要大几十倍至上百倍，横向干缩中弦向是径向的 2 倍。木材体积干缩率为最大，约等于 3 个方向干缩率之和。

（2）木材干缩系数

为了能比较不同含水率区段的干缩值，采用**干缩系数**（coefficient of shrinkage）这一指标，以计算确定出木材加工过程中板材尺寸应留出的干缩余量。

干缩系数是指吸着水每变化 1% 时木材的干缩率变化值，用 K 表示。弦向、径向、纵向和体积干缩系数分别记为 K_t、K_r、K_l 和 K_v，其值用木材的干缩率和引起干缩率的含水率差之比值表示：

$$K_{t,\ r,\ l,\ v} = \beta_w / (MC_1 - MC_2)$$

式中：β_w——气干状态下木材的弦向、径向、纵向、体积干缩率，%；

MC_1、MC_2——木材两个状态下的含水率，%；

木材干缩起点为纤维饱和点，一般以 30% 计算。利用干缩系数可算出纤维饱和点以下任何含水率时的干缩率：

$$\beta_w = K (30 - 100MC) \%$$

为了比较横向中径向和弦向干缩差异程度，常用差异干缩（differential shrinkage）来表示。

$$D = \beta_t / \beta_r = K_t / K_r$$

差异干缩是反映木材干燥时，是否易翘曲和开裂的指标。根据木材差异干缩的大小，大致可决定木材对特殊用材的适应性。表 5.1 列举了若干树种横向（径向、弦向）干缩率。

表 5.1 木材从生材状态到绝干状态的干缩率

树种	干缩率（%）			产地
	径向	弦向	体积	
杉木	1.23	2.91	4.20	中国
红松	1.22	3.21	4.59	中国
马尾松	1.50	2.96	4.66	中国
泡桐	1.47	2.69	4.53	中国
水曲柳	1.97	3.53	5.77	中国
白蜡	4.9	7.8	13.3	美国
山杨	3.5	6.7	11.5	美国
糖枫	4.8	9.9	14.7	美国
北部红橡	4.0	8.6	13.7	美国
黑胡桃	5.5	7.8	12.8	美国
美国铁杉	4.2	7.8	12.4	美国
火炬松	4.8	7.4	12.3	美国

（3）干缩带来的后果与控制措施

如图 5.6（a）所示，木材失去水分，会导致干缩；而木材吸收水分，会导致湿胀。这样会造成木制品在使用过程中出现尺寸变化，如抽屉在雨季难以关合，或在旱季出现较大裂缝；或者木地板在潮湿环境中出现起拱。此外，木板在失去水分时，由于内部含水率不均匀，造成内应力，若内应力大于木材横纹强度，就会出现木材开裂，如图 5.6（b）所示。

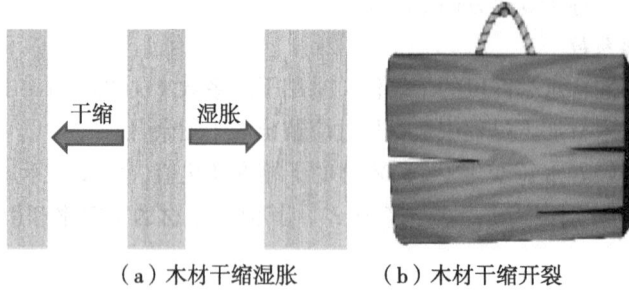

干缩   湿胀

（a）木材干缩湿胀   （b）木材干缩开裂

图 5.6　木材水分变化及造成的后果

木材干缩的控制措施有多种，具体介绍如下。

①表面涂饰　木材表面涂饰，可以形成一层具有保护性能的涂膜，增加木制品表面的美观性，同时可以防水、防湿，提高其尺寸稳定性（图 5.7）。

防水处理：在木材表面涂饰防水剂，隔绝木材与周围环境，从而提升木材的防水性。防水剂主要有硅油、石蜡、硅树脂、亚麻油等。

防湿处理：在木材表面涂饰或贴面，延缓湿空气在木材中的扩散速度，减少木材对水蒸气的吸着速率，减少膨胀和表面开裂的速度。常用的外表面覆面涂饰材料有油脂漆、天然树脂漆、酚醛树脂漆、氨基树脂漆、聚酯漆等。

图 5.7　木材涂饰

②高温干燥及热处理

高温干燥：将木材干燥至平衡含水率以下，利用吸湿滞后特性，减少木材尺寸变化。

热处理：木材热处理是以蒸汽、空气或者氮气等气体或植物油为传热介质，在一定温度（150~260℃）下进行加热处理，得到热处理材，称为"**炭化木**"。木材在该温度区间内，发生预炭化过程，主要是木材中的半纤维素发生降解，少量的纤维素和木质素也发生降解。按照传热介质，热处理可分为气相介质（蒸汽、空气）和液相介质（植物油）。

③木材乙酰化处理　乙酰化处理可以有效地将木材中的游离羟基转变成乙酰基，木材通过和醋酸酐（其稀释后就是醋酸）发生化学反应得以实现此结果（图 5.8）。从游离羟基到乙酰基的转变，大大降低了木材的吸水性，同时增强其尺寸稳定性和超强的耐久性。

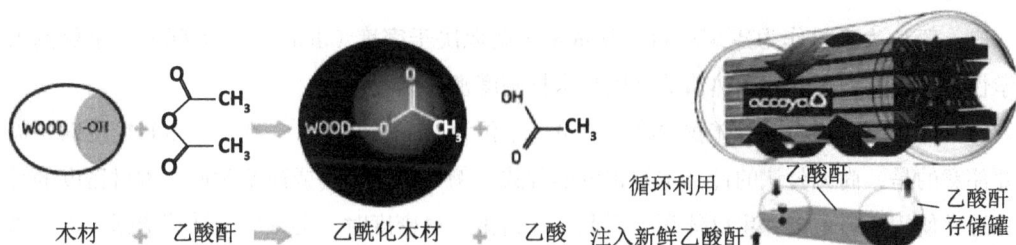

图 5.8　木材乙酰化处理（Diamond wood）

④正交组坯　将木材单板或锯材正交胶压，可利用干缩极小的纵向机械抑制横向干缩，将胀缩减小到最小。同时木材横纹方向强度小，而木材顺纹方向强度高，可以弥补木材横纹方向强度小的缺点。胶合板和正交胶合木的制造和利用就是基于此原理（图 5.9）。

图 5.9　胶合板和正交胶合木的正交结构

⑤利用指接材　将容易变形开裂的木材锯裁成小木方或利用小径材加工成木方、板材。通过有缝或无缝指接方式胶合成大木方或大板材，基本上可消除了木材变形和开裂的缺陷（图 5.10）。

⑥酚醛树脂处理　酚醛树脂处理是将低分子量的酚醛树脂注入木材，发生缩聚反应，生成不溶性树脂，提高木材尺寸稳定性。尽管低分子量酚醛树脂不和木材化学成分发生明显化学反应，但会进入结构松散的细胞壁非结晶区，形成物理填充。随着木

图 5.10　木材指接

材中树脂含量的增加，其抗胀缩率增加，当树脂含量达到 30%~40% 时，抗胀缩率增加迟缓。尺寸稳定性与酚醛树脂中的羟甲基酚的含量密切相关，因为羟甲基酚易与木材中的羟基形成氢键结合，减小木材的吸湿性。

⑦聚乙二醇处理　浸渍和涂饰聚乙二醇可显著减少木材的干缩湿胀，减少木材开裂、翘曲和变形，是一种常用的木材增容处理剂。聚乙二醇是由环氧乙烷与水（或乙二醇）发生加成反应而得到的链状聚合物，分子式为 $HO \cdot CH_2 - (CH_2 \cdot O \cdot CH_2)_n - CH_2OH$，相对分子质量为 1000 的聚乙二醇最适用于木材，几乎可以完全置换木材中的水分，分子量超过 3500 的聚乙二醇则不易进入木材的细胞壁。聚乙二醇一般通过常压浸渍渗入木材，聚乙二醇的水溶液浓度为 20%~25%，浸渍温度为 80℃，时间为 15~200min。

5.2　木材密度和孔隙度

5.2.1　木材密度含义

一般认为木材作为建筑材料，其重量主要取决于**密度**（density）。实际上，木材的力学性能、硬度、耐磨性和热值等性质与木材密度密切相关。

密度可以描述为单位体积的物体重量，单位为 g/cm³ 或 kg/m³。在讨论木材密度时，需要注意的是，尚无公认的计算木材密度的程式。例如，在运输或建筑领域，木材密度的计算通常使用生材重量和生材体积。因此，在讨论木材密度时，必须确定计算依据，即含水率。通过在同一含水率条件下去测定重量和体积从而计算密度，是一个很好的方法。

木材密度 = 重量 / 体积（在一定含水率条件下）

5.2.2　抽提物对木材密度的影响

木材通常含有数量可测的**抽提物**（extractive），包括萜烯、树脂、多酚（如单宁）、糖和油以及无机化合物，如硅酸盐、碳酸盐和磷酸盐。这些物质在心材中的含量高于边材。因而，心材的密度通常略高于边材。

木材中抽提物含量有多有少，从不到木材绝干重量的 3% 到超过 30% 不等。显然，这些物质的存在会对密度产生重大影响。

5.2.3　细胞壁密度和孔隙度

因为木材是多孔体，所以测得木材试样的体积是外形体积，即表观体积，这样的木材密度本质上是容积重或表观密度。如图 5.11 所示，固体中不含有任何孔隙，其外观体积称为实质体积。固体中含有孔隙，其尺寸远小于固体尺寸，其外观体积称为表观体积。散粒材料或粉状材料，在自然堆积状态下形成的体积，称为堆积体积。

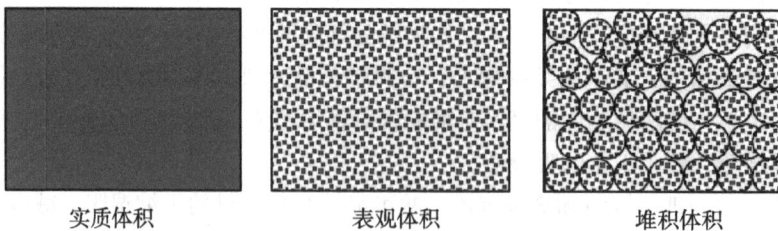

实质体积　　　　　表观体积　　　　　堆积体积

图 5.11　材料体积

木材是一种多孔体，其**实质密度**指的是木材细胞壁物质的密度，范围在 1.46~1.56g/cm³。本教材取 1.50g/cm³。所有树种的实质密度大致相同。可为什么不同树种的木材密度存在较大差异呢？这是因为木材孔隙[1]的影响。木材的**孔隙度**（porosity）可以用下式计算：

$$P = \left(1 - \frac{\rho_0}{\rho_{0w}}\right) \times 100\%$$

式中：P——木材孔隙度，%；

ρ_0——木材的绝干密度，g/cm^3；

ρ_{0w}——绝干木材的实质密度，g/cm^3；一般取平均值 1.50g/cm^3。

5.2.4　木材重量的计算

如果含水率和密度已知，木材的重量很容易求得。用木材的绝干密度和绝对含水率，计算木材重量的算法如下：

$$重量 = 体积 \times 木材绝干密度 \times （1 + MC）$$

5.2.5　四种木材密度

因为木材的体积和重量随含水率的变化而变化，因此木材的密度和含水率状态密切相关。如生材状态、气干状态、绝干状态的密度，分别称为生材、气干、绝干密度。此外，为了比较各树种密度，还引入了一个不随含水率变化的恒定密度的概念，即基本密度。

①基本密度　绝干材重量除以饱水（生材）的木材体积即为**基本密度**（basic density）。

$$\rho_b = \frac{m_0}{v_g}$$

式中：ρ_b——试样的基本密度，g/cm^3；

m_0——绝干材重量，g；

v_g——生材体积，cm^3。

②生材密度　生材重量除以生材体积即为生材密度（green density）。

$$\rho_g = \frac{m_g}{v_g}$$

式中：ρ_g——试样的生材密度，g/cm^3；

m_g——生材重量，g；

v_g——生材体积，cm^3。

③气干密度　气干材重量除以气干材体积即为气干密度（air-dry density）。

$$\rho_a = \frac{m_a}{v_a}$$

式中：ρ_a——试样的气干密度，g/cm^3；

---

[1]　区分两个概念：孔隙和空隙。孔隙指的是材料中未被密实固体充满的部分；而空隙指的是组成材料的各单元之间未被充满的部分，也叫间隙。

m_a——气干材重量，g；

v_a——气干材体积，cm^3。

由于各地平衡含水率不同，为了方便不同树种间进行比较，需将含水率调整到统一的状态。通常，将测得的木材气干密度，均换算成含水率为 12% 时的值。换算公式为：

$$\rho_{12} = \rho_a \times [1 - (1-K_v) \times (MC-12\%)]$$

式中：ρ_{12}——含水率为 12% 时的木材气干密度，g/cm^3；

ρ_a——含水率不是 12% 的木材气干密度，g/cm^3；

K_v——木材体积干缩系数；

MC——木材含水率，%。

④绝干密度　绝干材重量除以绝干材体积即为绝干密度（oven-dry density）。

$$\rho_o = \frac{m_o}{v_o}$$

式中：ρ_o——试样的绝干密度，g/cm^3；

m_o——绝干材重量，g；

v_o——绝干材体积，cm^3。

通常，基本密度 < 绝干密度 < 气干密度 < 生材密度。

5.2.6　木材密度的测定

（1）直接测量法

将试样加工成尺寸为 20mm×20mm×20mm 的标准立方体，相邻面要互相垂直。在试样各相对面的中心线位置画圈，用螺旋测微仪分别测出其径向、弦向和顺纹方向的尺寸，准确至 0.01mm，用千分之一的天平称重，准确至 0.001g。

气干密度试样以气干材制作，测量气干尺寸后立即称气干重，然后放入烘箱，用烘干法测出试样的绝干重量。试样烘干后，可立即测出绝干状态下体积。按上述公式计算气干密度和绝干密度。

（2）排水法[①]

利用水的密度为 $1g/cm^3$，试样入水后排出水的重量，与试样体积数值相等的原理而设计的（图 5.12）。测定时，将盛有一定深度水的烧杯放置在电子天平上，金属针浸入水下 1~2cm 后，天平读数；然后将金属针尖插固于已称重的试样上并浸入水中，再次读数。两次读数之差，即为试样的体积。该方法对形状不规则的试样适合，尤其适合测定饱水状态下试样的体积。

（3）快速测定法

将试样制成 2cm×2cm×20cm 的长方体，要求试样平直、规整，上下两端面相互平行。把试样全长刻划区分成 10 等分，依次标记为 0.1、0.2、0.3、…、0.9。然后将试样标

[①]　在用排水法或快速测定法测定干木材试件时，表面要涂一层石蜡，以防止应木材吸收水分而引起的体积测量误差。

记 0.1 的一端浸入盛有水的玻璃筒中，且不与量筒内壁接触，此时在水面处的试样标记，就为该木材的密度（图 5.13）。

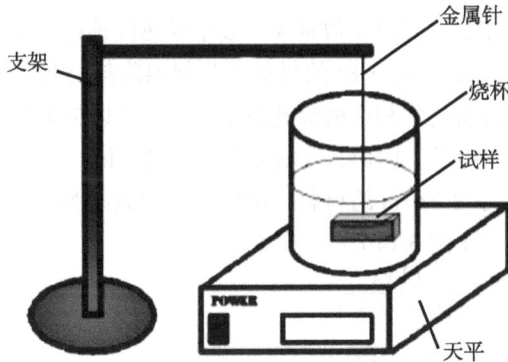

图 5.12　排水法测定浸渍体积和生材体积　　图 5.13　利用水位差快速测定木材体积

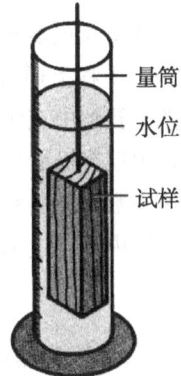

5.3　木材的其他物理性质

5.3.1　木材的热学性质

（1）比热

某物质平均温度升高 1 单位所需的热量称为该物质的热容。物质单位质量的热容与相同质量水的热容之比称为**比热容**（specific heat），简称比热。比热可以简化为单位质量的物质在温度升高或降低 1 单位时所吸收或放出的热量。1913 年，杜拉普（F. Dunlap）利用热量计法测定了 20 个树种的 100 个试样在 0~106℃之间的比热，推算出绝干木材比热的经验公式[①]：

$$c_0 = 1.112 + 0.00485T$$

式中：c_0——绝干材的比热，kJ（kg·K）；

　　　T——开尔文温度，K。

若将湿木材看作是木材、水和空气所组成的三相系统，按热容叠加原理并略去空气的热容，利用湿木材热容量等于水的热容与绝干材热容量之和的等式关系，可得出湿木材的比热：

$$c_{MC} = \frac{(1+MC)m_0 - m_0}{(1+MC)m_0}c_w + \frac{m_0}{(1+MC)m_0}c_0 = \frac{MC \cdot c_w + c_0}{1+MC}$$

式中：c_{MC}——湿材的比热，kJ/（kg·K）；

　　　c_0——绝干材的比热，kJ/（kg·K）；

　　　MC——木材的绝对含水率，%；

　　　m_0——绝干材的重量，kg；

---

① 1cal=4.18J。

c_w——水的比热，kJ/（kg·K）。

（2）木材导热系数

导热系数（thermal conductivity）是建筑材料最重要的热湿物性参数之一，与建筑能耗、室内环境及很多其他热湿过程息息相关。导热系数越大，表示导热性越好；反之，导热性能差。工程上通常将导热系数小于 0.23W/(m·K) 的材料作为保温隔热材料。木材具有多孔性，孔隙中充满空气，由于木材细胞之间相通的纹孔较小，空气难以自由对流，再加上木材中仅有极少易于传递能量的自由电子。因而，木材是热的不良导体。

导热系数是指在稳定传热条件下，厚度为 1m 的材料，两侧表面的温差为 1K 或 1℃，在一定时间内，通过 $1m^2$ 面积传递的热量（图 5.14）。

$$\lambda = \frac{Q\Delta x}{At(T_1 - T_2)}$$

式中：λ——导热系数，W/(m·K) 或 W/(m·℃)；

　　　Q——传导的热量，J；

　　　Δx——木材的厚度，m；

　　　A——木材的传热面积，m^2；

　　　$T_1 - T_2$——木材两侧的温度差，℃ 或 K；

　　　t——传热时间，s。

木材导热系数也呈现明显的各向异性（图 5.15）。纵向（顺纹）导热系数大。横向中径向由于木射线组织的存在，其导热系数比弦向大 5%~10%。在实际使用中，常以径、弦向的平均值作为横向导热系数的数值。

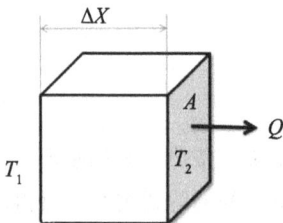

图 5.14　木材热传递示意图

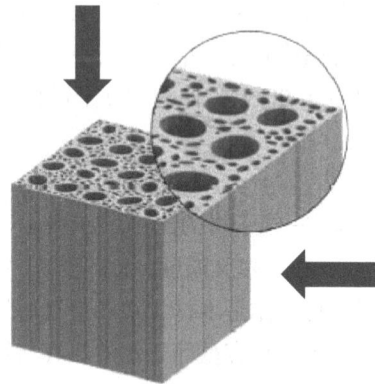

图 5.15　木材多孔性与热流方向

木材导热系数随着木材密度的增加而成比例增加，几乎成直线关系。绝干木材横向导热系数与密度间的关系大致符合以下公式：

$$\lambda = (5.18\rho_0 + 0.57P) \times 418 \times 10^{-4} = 0.217\rho_0 + 0.0238P$$

式中：λ——绝干木材横纹导热系数，W/（m·K）；

　　　ρ_0——木材绝干密度，g/cm^3；

　　　P——木材孔隙度，%。

上式实际上是由木材实质与空气之迭加。若 $\rho_0 \to 0$，则 $P \to 100\%$，$\lambda = 0.0238\text{W/}（\text{m}\cdot\text{K}）$，即为干空气的导热系数；若 $P \to 0\%$，则 $\rho_0 \to 1.50$，$\lambda = 0.325\text{W/}（\text{m}\cdot\text{K}）$，即为木材细胞壁实质的导热系数。因此，木材的密度越小，孔隙率就越大，则导热系数越小，绝热性越好。锅盖习惯用轻软的杉木和泡桐等木材制作的原因就在于此。表 5.2 为几种常见建材的导热系数。

表 5.2　几种常见建材的导热系数

材料	导热系数 [W/（m·K）]	材料	导热系数 [W/（m·K）]
结构用针叶树材 （含水率 12%，横向）	0.10~0.14	混凝土	0.9
		矿物棉	0.036
铝	216	玻璃	1.0
钢	45	石膏	0.7

水的导热系数约为空气的 25 倍，随着木材含水率的增加，部分空气被水分替代，因而木材的导热系数将增加。

（3）木材的导温系数

导温系数又称**热扩散系数**（thermal diffusivity），表征的是材料在加热或冷却非稳态传热过程中各点温度趋于一致的能力。导温系数大，说明木材在加热或冷却过程中温度上升或下降快。表 5.3 为常见建材的热工参数。

表 5.3　常见三种建筑材料的热工性能

材料	比热 c[kJ/（kg·K）]	导热系数 λ[W/（m·K）]	导温系数 α（m²/s）
松木	2.51	0.16（横纹）	0.11×10^{-6}
		0.35（顺纹）	0.24×10^{-6}
钢	0.47	58	16.16×10^{-6}
普通混凝土	0.88	1.60	0.71×10^{-6}

木材的导温系数（α）与导热系数（λ）、比热（c）和密度（ρ）之间的关系如下：

$$\alpha = \lambda / c\rho \quad（\text{m}^2/\text{s}）$$

5.3.2　木材的导电性

木材在气干状态下，其导电性是弱小的，特别是绝干材可视为绝缘体，因此木材可作为交通、电力及其他行业上重要的绝缘材料之一。但如果木材中含有水分，特别是在纤维饱和点以下，含水率越高，木材导电性越强。生材为电的导体，所以雨中树木易被雷电击倒。下雷阵雨时，不要去树下躲雨（图 5.16）。

图 5.16　湿木材导电性

▶ **思考与训练**

1. 名词解释。

纤维饱和点、平衡含水率、吸湿滞后、木材密度、差异干缩。

2. 简述木材细胞壁中水分存在的形式及其对材性的影响。

3. 简述木材各向干缩差异现象及其发生的原因。

4. 如何控制木材干缩湿胀？

5. 某体育馆地板采用钉接木条地板，木条之间基本上没有裂缝或缝隙。体育馆的尺寸为 15.2m×36.6m，条形地板与体育馆长轴平行。地板采用的是含水率为 6% 的硬枫木（糖枫）。地板被钉在 38mm×89mm 的木钉条上。木钉条用胶泥粘在混凝土地面上。木条地板和混凝土砌块墙之间留有不到 25mm 的预留空间。问题在于木地板在混凝土地面完全干燥之前就已经钉好了。因此，枫木地板吸湿，含水率增加到 9% 左右。在地板安装后就出现了翘曲。问题在哪里？如何解决？

6. 木材密度种类及其测定方法有哪些？

7. 同一块木材的哪一种密度数值最小？为什么？

8. 为什么说纤维饱和点是木材各种性质的转折点？

9. 描述吸湿滞后现象及其生产应用。

第 **6** 章
木材力学性质

▶ **学习目标**

（1）掌握木材力学基础知识；

（2）了解木材关键力学性质的概念与测定方法。

▶ **本章描述**

　　木材作为生物质材料，具有独特的构造特征和生物学特征，这些特征赋予了木材特殊的力学性质。木材各向异性在力学性质上如何体现？有哪些主要力学性质？本章将解答这些疑问。

　　木材作为一种非均质、各向异性的天然高分子材料，许多性质都有别于其他材料，而其力学性质和其他均质材料有着明显的差异。木材力学性质包括应力与应变、弹性、粘弹性、强度、硬度、抗劈力以及耐磨性等。

6.1　木材力学基础

6.1.1　应力 – 应变关系

　　构件因受到外力的作用而变形，其内部各部分之间的相互作用力也发生改变。这种由于外力作用而引起的构件内部各部分之间的相互作用力的改变量，称为**内力**。应力是指物体单位截面上的内力（图 6.1）。**变形**（deformation）是指构件在外力作用下，几何形状和尺寸的改变。按照木材构件的基本变形形式，可分为**压缩**（compression）、**拉伸**（tension）、**弯曲**（bending）、**剪切**（shear）和**扭转**（twisting）（图 6.2）。其中，轴向拉伸和压缩产生拉应力和压应力，弯曲产生拉应力、压应力和剪应力，剪切和扭转产生剪应力。

　　（1）**拉应力**

　　两个大小相等而方向相反的外力沿着木材同一方向线作用，引起木材拉伸变形，此时垂直于木材构件轴线的截面上的应力称为**拉应力**（tensile stress）。

（2）**压应力**

两个大小相等而方向相反而相对的外力，沿着木材同一方向作用引起木材压缩变形。此时垂直于木材构件轴线的截面上的应力称为**压应力**（compressive stress）。

（3）**剪应力**

两个大小相等方向相反接近平行外力作用于木材，促使木材一部分相对于另一部分发生错动的剪开现象，此时错开面上产生应力称为**剪应力**（shear stress），用 τ 表示。

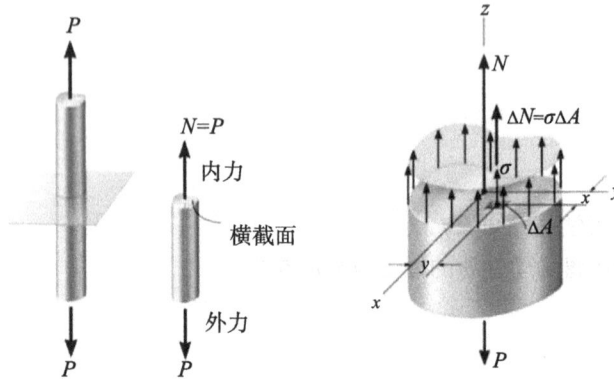

图 6.1　内力与应力

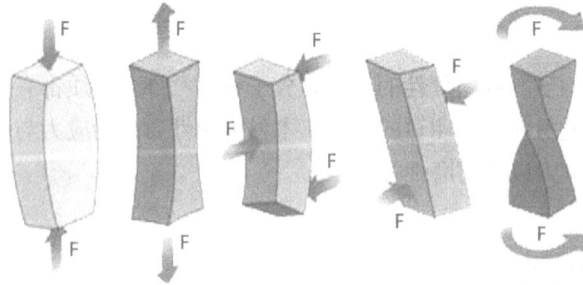

图 6.2　木材构件的基本变形形式

一般构件内各点处的变形是不均匀的。将构件在单位长度上所产生的线应变，称为**线应变**（linear strain），用 ε 表示。构件中两个相互垂直的面在受力变形后以弧度表示的夹角的改变量，称为**剪应变**（shear strain），用 γ 表示。

物体在外力作用下会发生变形（图 6.3）。与之对应的是，应力会导致物体产生应变。外力在一定限度内，应力和应变成正比的线性关系；外力超过这一限度，二者直线关系变为曲线关系。随着外力增大，物体最终出现破坏，此时应力达到最大值。以应力为纵坐标，应变为横坐标，表示应力和应变关系的曲线称为应力 – 应变曲线。

图 6.4 为木材顺纹压缩应力和应变关系曲线。在比例极限以下，应力和应变的比值是一个常数，应力与应变成正比例直线关系的顶点为比例极限。超过比例极限后，应力与应变关系不再是直线，但变形仍然是弹性的，即解除应力后变形完全消失，此时的应力是材料弹性变形的极限值。弹性范围内的变形称之为弹性变形。木材的弹性极限和比例极限相

差不大，不做区分。随着外力增大，应力超过弹性极限，此时外力去除后，材料变形只能部分恢复，残留的部分变形称之为塑性变形。

图 6.3　物体受力变形

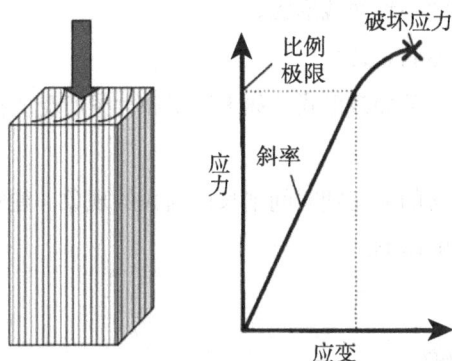

图 6.4　木材顺纹压缩应力 – 应变曲线

在比例极限以下，木材应力和应变存在下列关系，即**胡克定律**：

$$\sigma = E\varepsilon$$

式中：E——**弹性模量**，MPa 或 N/mm^2。

弹性模量为物体在外力作用下抵抗变形的能力，它是材料刚性的指标，简称模量。

将木材拉伸和压缩试验中应力和应变的比值称为杨氏模量，与木材弯曲时的弯曲模量加以区分。实际上，木材的拉伸、压缩和弯曲模量大致相等，但压缩的弹性极限比拉伸弹性极限低得多。

同理，当剪应力未超过某一极限值时，剪应力与其相应的切应变成正比。引入比例常数，则可得到，

$$\tau = G\gamma$$

上式称为**剪切胡克定律**（Hooke's law）。式中的比例常数 G 称为材料的剪切模量。它也与材料的力学性能有关。对同一材料，剪切模量 G 为常数，可由试验测定。G 的单位与应力的单位相同。

6.1.2 木材力学性质的若干概念

（1）**强度**（strength）

木材抵抗外力破坏的能力。

（2）**刚度**（stiffness）

木材抵抗外力造成尺寸和形状变化的能力。

（3）**韧性**（toughness）

木材吸收能量和抵抗反复冲击荷载或抵抗超过比例极限的短期应力的能力。

（4）**硬度**（hardness）

木材抵抗其他刚性物体压入的能力。

（5）**静力荷载**（static load）

荷载是缓慢的、均一的一种加载形式。

（6）**长期荷载**（long-term load）

力作用时间相当长的一种加载形式。如木屋架和大梁所受的荷载形式。

（7）**蠕变**（creep）

在恒定的应力下，木材的应变随时间增长而增大的现象。建筑木构件在长期承受静力荷载时，要考虑蠕变带来的影响。

6.2 木材弹性常数

木材的各向异性取决于组织排列。木材的轴向细胞是平行于树干延伸的，而射线细胞垂直于树干径向延伸，另组成细胞壁的各层微纤丝排列不同，以及微晶为单斜晶体，加上木材的层次性必然使木材显示出各向异性。

普莱斯首先把正交对称原理应用于木材，借以说明木材弹性的各向异性。如图6.5所示，按木材径面、弦面取一个立方体，试件有3个对称轴，平行于纵向取L轴，平行于径向取R轴，平行于弦向取T轴。如把这三轴近似看作相互垂直的弹性对称轴，就可以用正交对称原理来讨论。符合正交对称性的材料有9个独立弹性常数。

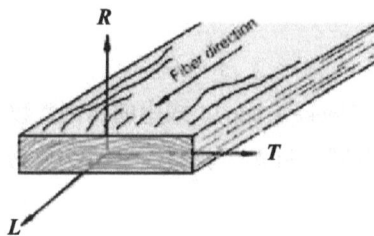

图6.5 木材正交对称性

表 6.1　典型针叶树材和阔叶树材的弹性常数

材料	密度（g/cm³）	含水率（%）	E_L (MPa)	E_R (MPa)	E_T (MPa)	G_{LT} (MPa)	G_{LR} (MPa)	G_{TR} (MPa)	μ_{RT}	μ_{LR}	μ_{LT}
针叶树材											
云杉	0.390	12	11583	896	496	690	758	39	0.43	0.37	0.47
花旗松	0.590	9	16400	1300	900	910	1180	79	0.63	0.43	0.37
阔叶树材											
轻木	0.200	9	6274	296	103	200	310	33	0.66	0.23	0.49
山毛榉	0.750	11	13700	2240	1140	1060	1610	460	0.75	0.45	0.51

根据表 6.1 中所列的木材的弹性常数得出以下结论：

①木材是高度异向性材料，弹性模量 E 在 3 个方向各不相同。纵向弹性模量大于横向十几倍甚至几十倍，横向中径向大于弦向。

②剪切模量 G，木材横断面最小，径、弦面剪切模量分别与径、弦向弹性模量数值相近。

③弹性模量 E 和剪切模量 G 均随密度增加而增加。

④木材的泊松比均小于 1。

6.3　木材主要力学性能指标

6.3.1　木材的抗拉强度

外力作用于木材，使其发生拉伸变形，**木材这种抵抗拉伸破坏的最大能力，称为抗拉强度**。视外力作用于木材纹理的方向，**木材抗拉强度分为顺纹抗拉强度和横纹抗拉强度**。

木材顺纹抗拉强度，是指木材沿纹理方向的截面承受的最大拉力荷载。木材顺纹抗拉强度是木材的最大强度，木材在使用中很少出现因拉断而破坏。

木材横纹抗拉强度，是指垂直于木材纹理方向的截面承受的最大拉力荷载。木材的横纹抗拉强度比顺纹抗拉强度低得多，一般只有顺纹抗拉强度的 1/40~1/30。

根据国家标准《木材顺纹抗拉强度试验方法》（GB/T 1938—2009）规定，试样的形状和尺寸如图 6.6 所示。试验以均匀速度加载，在 1.5~2.0min 内使试样破坏，破坏荷载精确至 100N。根据国家标准《木材横纹抗拉强度试验方法》（GB/T 14017—2009）规定，试样的形状和尺寸如图 6.6 所示。试验以均匀速度加载，在 1.5~2.0min 内使试样破坏，破坏荷载精确至 10N。

木材抗拉强度按下式计算：

$$\sigma_W = \frac{P_{max}}{bt}$$

式中：σ_W——试样含水率为 W 时的顺纹抗拉强度，MPa；

　　　　P_{max}——破坏荷载，N；

b ——试样宽度，mm；

t ——试样厚度，mm。

试样含水率为 12% 时的阔叶树材的顺纹抗拉强度，应按下式计算：

$$\sigma_{12} = \sigma_W \left[1 + 0.015(W - 12) \right]$$

若试样为针叶树材，含水率在 9%~15% 时，取 $\sigma_{12} = \sigma_W$。

试样含水率为 12% 时的横纹抗拉强度，应按下式计算。

径向试样为：$\sigma_{12} = \sigma_W \left[1 + 0.01(W - 12) \right]$；

弦向试样为：$\sigma_{12} = \sigma_W \left[1 + 0.025(W - 12) \right]$。

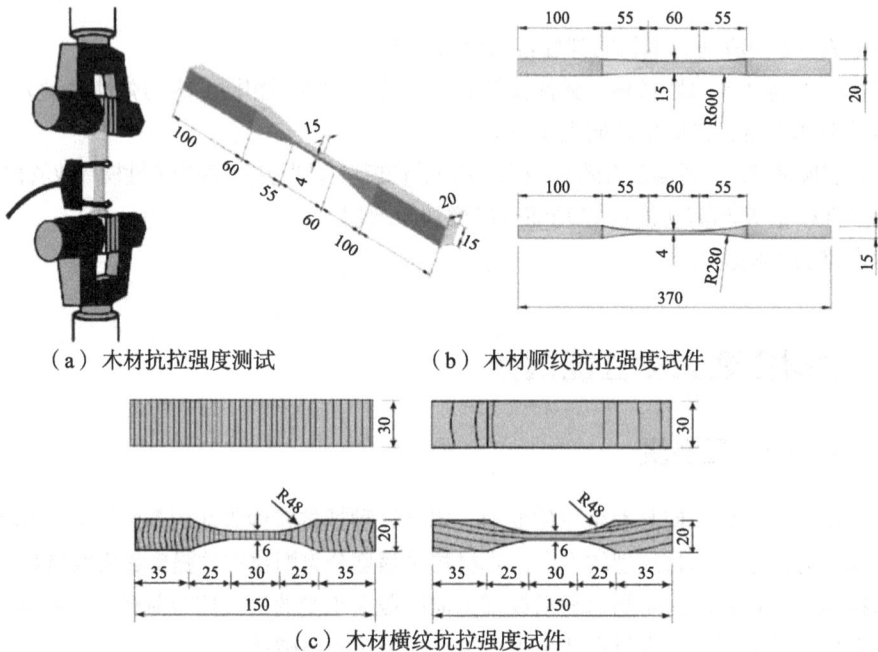

（a）木材抗拉强度测试　　　（b）木材顺纹抗拉强度试件

（c）木材横纹抗拉强度试件

图 6.6　木材抗拉强度测试示意图

6.3.2　木材的抗压强度

木材抗压强度分为**顺纹抗压强度**和**横纹抗压强度**。

根据我国国家标准《木材顺纹抗压强度试验方法》（GB/T 1935—2009），木材顺纹抗压强度的测试原理是沿木材顺纹方向以均匀速度施加压力至破坏，以确定木材的顺纹抗压强度。试样尺寸为 30mm×20mm×20mm，长度为顺纹方向（图 6.7）。试验含水率为 W（%）时的顺纹抗压强度，应按下式计算：

$$\sigma_W = \frac{P_{\max}}{bt}$$

式中：σ_W ——试样含水率为 W 时的顺纹抗压强度，MPa；

$P_{\max}$ ——破坏荷载，N；

b ——试样宽度，mm；

t ——试样厚度，mm。

试样含水率为 12% 时的顺纹抗压强度，按下式计算：

$$\sigma_{12} = \sigma_W\left[1+0.05\left(W-12\right)\right]$$

我国木材顺纹抗压强度的平均值约为 45MPa。顺纹比例极限与强度的比值约为 0.7，针叶树材约为 0.78，软阔叶树材约为 0.7，硬阔叶树材约为 0.66。

木材横纹抗压强度指垂直于木材纹理方向承受压力荷载的比例极限应力。木材横纹抗压强度的测定有两种方式：横纹全部抗压强度和横纹局部抗压强度（图 6.8）。木材横纹全部抗压试验的试样尺寸为 30mm × 20mm × 20mm，长度为顺纹方向。木材横纹局部抗压试验的试样尺寸为 60mm × 20mm × 20mm，长度为顺纹方向；在试样中部加压钢块尺寸为 30mm × 20mm × 10mm。

横纹全部抗压　　　横纹局部抗压

图 6.7　木材顺纹抗压强度测试　　　图 6.8　木材横纹抗压强度测定试样与受力方向

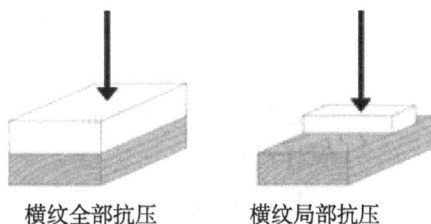

试样含水率为 W（%）时的径向或弦向的横纹全部或局部抗压比例极限应力，按下式计算：

$$\sigma_{yW} = \frac{P}{ab}$$

式中：σ_{yW} ——试样含水率为 W 时的径向或弦向的横纹局部抗压比例极限应力，MPa；

P ——比例极限荷载，N；

a ——试样长度或加压钢块宽度，mm；

b ——试样宽度，mm。

试样含水率为 12% 时径向或弦向的横纹全部或局部抗压比例极限应力，应按下式计算：

$$\sigma_{y12} = \sigma_{yW}\left[1+0.045\left(W-12\right)\right]$$

6.3.3　木材抗弯弹性模量和抗弯强度（三点弯曲）

抗弯强度只做弦向试验，可以采用三点弯曲（图 6.9）。同样，试验也可以得到木材的抗弯弹性模量。根据国家标准，抗弯弹性模量测定与抗弯强度的试验采用同一试样。

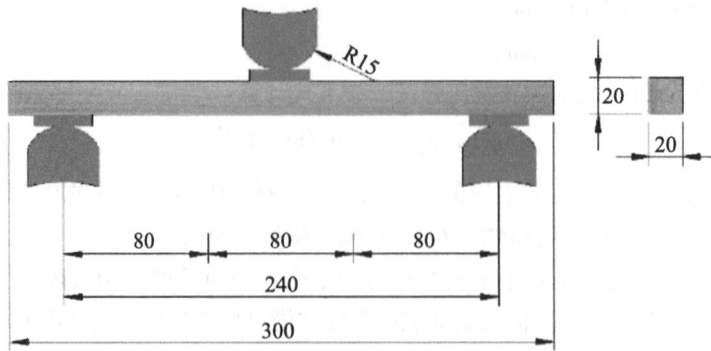

图 6.9　木材抗弯弹性模量和抗弯强度的测定（三点弯曲）

试样含水率为 W（%）时的抗弯弹性模量按下式计算：

$$E_W = \frac{Pl^3}{4fbh^3}$$

式中：P——为集中荷载，N；

l——两支座间跨距，240mm；

b——试样宽度，mm；

h——试样高度，mm；

f——试件中点底面挠度，mm。

试样含水率为 W 时的抗弯强度按下式计算：

$$\sigma_W = \frac{3P_{max}l}{2bh^2}$$

式中：σ_W——试样含水率为 W 时的抗弯强度，MPa；

P_{max}——破坏荷载，N；

l——两支座间跨距，mm；

b——试样宽度，mm；

h——试样高度，mm。

6.3.4　木材抗弯弹性模量和抗弯强度（四点弯曲）

理论上，四点弯曲测得的木材抗弯弹性模量和抗弯强度更为准确（图 6.10）。在木梁跨高比 [①] 较大时，三点弯曲和四点弯曲测得的木材抗弯弹性模量和抗弯强度差别不大。

试样含水率为 W（%）时的抗弯弹性模量按下式计算：

$$E_W = \frac{23Pl^3}{108fbh^3}$$

式中：P——集中荷载，N；

l——两支座间跨距，240mm；

b——试样宽度，mm；

---

① 跨高比＝跨度：试样高度。根据计算，跨高比小于 6，梁在三点弯曲荷载作用下会产生明显的剪切效应。

h ——试样高度，mm；

f ——试件中点底面挠度，mm。

试样含水率为 12% 时的抗弯弹性模量按下式计算：

$$E_{12} = E_W \left[1 + 0.015(W-12) \right]$$

若继续加载，直至试件破坏，可以得到木材的抗弯强度，计算如下：

$$\sigma_{bW} = \frac{P_{max} l}{b h^2}$$

式中：P_{max} ——为集中荷载，N；

l ——两支座间跨距，mm；

b ——试样宽度，mm；

h ——试样高度，mm。

试样含水率为 12% 时的抗弯强度按下式计算：

$$\sigma_{b12} = \sigma_{bW} \left[1 + 0.04(W-12) \right]$$

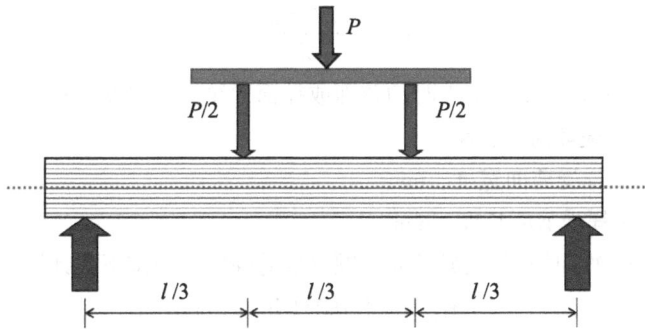

图 6.10　木材抗弯弹性模量和抗弯强度的测定（四点弯曲）

6.3.5　木材抗剪强度

木材抗剪强度视外力作用于木材纹理的方向，分为顺纹抗剪强度和横纹抗剪强度（图 6.11）。在实际应用中，发生横纹剪切的现象不仅罕见，而且横纹剪切总是要横向压坏纤维产生拉伸作用而非单纯的横纹剪切，因此不作为材性指标进行测定。木材的横纹剪切强度为顺纹剪切强度的 3~4 倍。

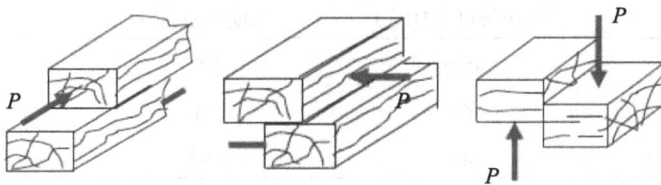

图 6.11　木材剪切示意图

木材顺纹抗剪试样形状与尺寸如图 6.12 所示，试样受剪面应为径面或弦面，长度为顺纹方向。试样缺角部分的角度为 $106°\ 40'$，允许误差为 $\pm 20'$。试验装置如图 6.13 所示。

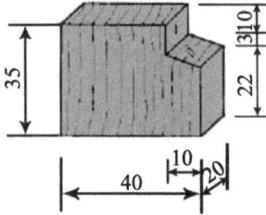

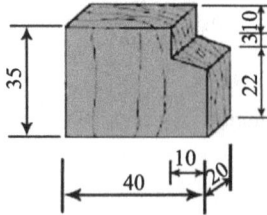

图 6.12　顺纹抗剪试样　　　　　　　　图 6.13　顺纹抗剪试验装置

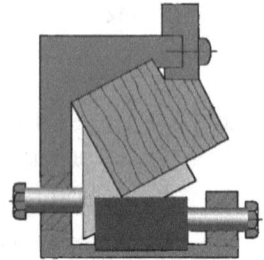

试样含水率为 W 时的弦面或径面顺纹抗剪强度，应按下式计算：

$$\tau_W = \frac{0.96 P_{max}}{bl}$$

式中：τ_W——试样含水率为 W 时的弦面或径面顺纹抗剪强度，MPa；

　　　　P_{max}——破坏荷载，N；

　　　　b ——试样受剪面宽度，mm；

　　　　l ——试样受剪面长度，mm。

试样含水率为 12% 时的弦面或径面的顺纹抗剪强度，应按下式计算：

$$\tau_{12} = \tau_W [1+0.03\,(W-12)]$$

木材的各强度之间关系和常见建筑材料的比强度见表 6.2、表 6.3。

表 6.2　木材的各强度之间关系（以顺纹抗压强度为 1）

抗压强度		抗拉强度		抗弯强度	抗剪强度		
顺纹抗压	横纹抗压	顺纹抗拉	横纹抗拉		顺纹剪切	横纹弦向剪切	横纹径向剪切
1	1/10~1/3	2~3	1/20~1/3	1.5~2	1/7~1/3	1/14~1/6	1/2~1

表 6.3　常见建筑材料的比强度（强重比）

材料	抗压强度（MPa）	密度（g/cm³）	比强度
低碳钢	400	7.8	51
混凝土	60	2.4	25
松木（顺纹）	45	0.55	82

6.4 木材力学性质的影响因素

6.4.1 木材的纤维组织

木材受力时，主要靠细胞壁承受外力，细胞组成的纤维组织[①]越均匀密实，强度就越高（图 6.14）。

厚壁细胞数量 ↑
细胞壁厚度 ↑
表观密度 ↑   木材的强度 ↑
晚材含量 ↑

图 6.14 木材构造对力学性质的影响

6.4.2 木材水分的影响

木材含水率对木材力学性质的影响，是指纤维饱和点以下木材水分变化时，给木材力学性质带来的影响。含水率在纤维饱和点以下，木材强度随着木材水分的减少而增大，随着水分的升高而降低，主要是由于单位体积内纤维素和木质素分子的数目增多，分子间的结合力增强所致。含水率高于纤维饱和点，自由水含量增加，其强度值不再减小，基本保持恒定。

如图 6.15 所示，木材含水率变化对不同强度的影响不同，对抗弯和顺纹抗压强度影响较大，对顺纹抗剪强度影响较小，对顺纹抗拉强度几乎没有影响。

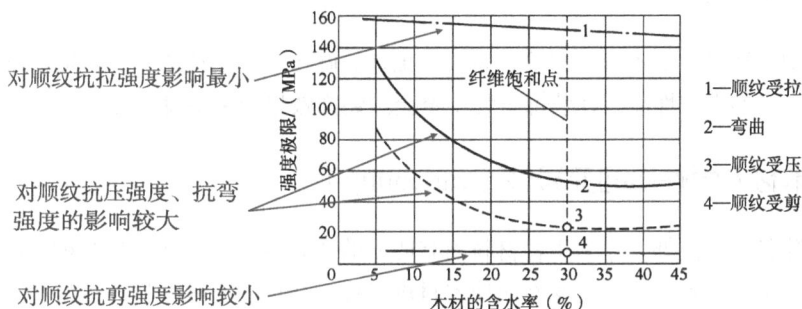

对顺纹抗拉强度影响最小

对顺纹抗压强度、抗弯强度的影响较大

对顺纹抗剪强度影响较小

纤维饱和点

1—顺纹受拉
2—弯曲
3—顺纹受压
4—顺纹受剪

图 6.15 含水率对木材强度的影响

我国木材试验标准规定，以标准含水率（即含水率 12%）时的强度为标准值，其他含水率时的强度，可按下式换算成标准含水率时的强度：

$$\sigma_{12} = \sigma_W [1 + \alpha (W-12)]$$

表 6.4 列出了各强度含水率调整系数值。

表 6.4 我国木材物理力学试验方法中各种强度的含水率调整系数值

强度性质	α 值	强度性质	α 值
顺纹抗拉	0.015	顺纹抗剪	0.03
抗弯	0.04	横纹抗压	0.045
顺纹弹性模量	0.015	横纹抗压弹性模量	0.055
顺纹抗压	0.05	硬度	0.03

注：顺纹抗拉含水率的调整只限于阔叶树材，针叶树材不进行调整。

[①] 纤维组织在针叶树材中主要指的是管胞，在阔叶树材中主要指的是木纤维。

6.4.3 荷载持续时间

木材对长期荷载的抵抗能力与对短期荷载不同。木材在外力作用下产生的等速蠕变，经过长时间作用后，会急剧产生大量的连续变形。木材在长期荷载作用下所能承受的最大应力称为木材的持久强度。木材的持久强度比其极限强度小得多，一般为极限强度的50%~60%。一切木结构都处于荷载的长期作用下。因此在设计木结构时，应考虑荷载时间对强度的影响，以持久强度作为设计依据。

6.4.4 温度

木材的强度随环境温度升高而降低。研究表明，处理温度在25~50℃上升时，针叶树材气干材抗拉强度降低10%~15%，抗压强度降低20%~24%。当木材长期处于60~100℃时，会引起水分和所含挥发物的蒸发而呈暗褐色，强度下降，变形增大。温度超过140℃，木材中的半纤维素先发生热裂解，强度明显下降。因此，长期处于高温的建筑物不宜采用木结构。

6.4.5 木材缺陷

木材中的缺陷，如木节、斜纹、裂纹、虫蛀、腐朽等，会造成木材构造的不连续性和不均匀性，影响力学性能，在装饰工程中更会给装饰效果带来不良的影响。

6.4.6 木材力学性质的变异性（variability）

木材除各向异性外，作为生物材料尚有明显的变异性，木材性质随树种、产地和部位等的不同而不同，木材具有变异性而使木材难以称为真正的工程材料，在材性测试时，需按标准取材并需变数统计[①]，增加了材性研究的复杂性。通过单元重组制造人造板材，可以有效降低木质材料的变异性（图 6.16）。

木材物理力学性质变数统计的主要指标有 5 项：

（1）算术平均值，又称均值，用 M 表示。

$$M = \frac{\sum\limits_{i=1}^{n} X_i}{n}$$

其中，n 表示变数的个数。

（2）均方差，又称标准差，用 σ 表示。均方差越大，说明数列离散程度越大。

$$\sigma = \sqrt{\frac{\sum\limits_{i=1}^{n}\left(X_i - M\right)^2}{n-1}}$$

---

① 变数统计，指的是不仅要算出一组数据的平均数，而且还要求出其变异系数及讨论可靠性等各项指标。

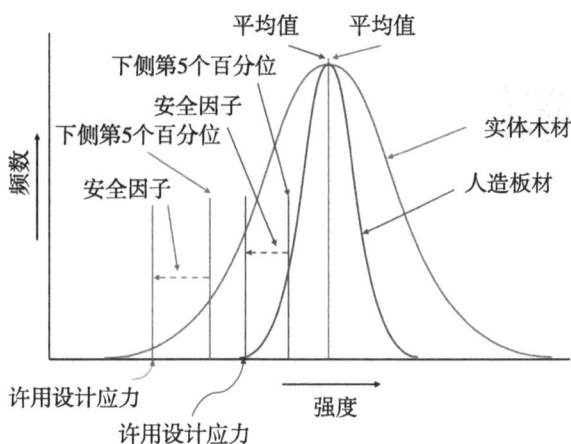

图 6.16　实体木材和人造板材的强度密度分布

（3）变异系数，用 V 表示。表征数列的相对离散程度。

$$V = \frac{\sigma}{M} \times 100\%$$

（4）均值误差，又称平均误差，用 m 表示，它是判断平均数可靠性的指标，m 越小越可靠。

$$m = \frac{\sigma}{\sqrt{n}}$$

（5）准确指数，用 P 表示，它是判别试验结果可靠性的指标。P 越小，试验结果越可靠。

$$P = \frac{m}{n} \times 100\%$$

根据森工试验只有准确指数 $P<5\%$，才能保证试验充分可靠性。同时它亦有确定试验试件数的重要指标。通常最少的试件数 n 可以通过变异系数与准确指数 P 比值的平方来确定。

$$n_{\min} = \left(\frac{V}{P}\right)^2$$

思考与训练

1.什么是内力？什么是应力？

2.影响木材力学性质的因素有哪些？

> **学习目标**
> （1）了解木材锯解的方法和优缺点；
> （2）了解锯材干燥的特点；
> （3）了解木质人造板的最新发展和特点。

> **本章描述**
> 　　木材如何加工？木质材料如何重组、复合，才能最大限度提升其物理力学性能？本章将解答这些疑问。

　　木材作为传统的材料，一直为人类所利用。与其他材料相比，木材具有多孔性、各向异性、湿胀干缩性、燃烧性和生物降解性等独特性质，如何更好地利用这些特性和最大限度地限制其副作用，是木材科学家和工程技术专家长期努力解决的主要问题。

　　随着自然资源和人类需求发生变化以及科学技术进步，木材利用方式从原始的原木逐渐发展到**锯材**（sawn lumber or sawn timber）、**单板**（veneer）、**刨花**（particle/flake/chip）、**纤维**（fiber）和化学成分利用，形成了一个庞大的新型木质材料家族，如**胶合板**（plywood）、**刨花板**（particleboard）、**纤维板**（fiberboard）、**单板层积材**（laminated veneer lumber）、**层板胶合木**（glue-laminated timber）、**正交胶合木**（cross-laminated timber，CLT）、**重组木**（scrimber）、**定向刨花板**（oriented strand board）、**重组装饰薄木**（reconstituted decorative veneer）等**木质重组材料**（Wood-based reconstituted materials），以及**石膏刨花板**（gypsum particleboard）、**水泥刨花板**（cement particleboard）、**木/塑复合材料**（wood-plastic composite，WPC）、**木材/金属复合材料**（wood-metal composite，WMC）、**木质导电材料**（wood conductive material）和**木材陶瓷**（wood ceramic）等**木质复合材料**（wood composite）。

7.1　木材锯解

　　一般伐木选择在冬季，因为树木在冬季长得慢。在西方，树木采取**皆伐**（clear felling/

cutting）方式（伐区内的成熟林木短时间内全部伐光或者几乎全部伐光）。在皆伐之后，树木被运到**锯木厂**（sawmill）切割成合适尺寸（图 7.1）。

图 7.1 树木皆伐、原木运输与制材

木材锯解（log conversion）有 3 种方式：

（1）**常规下锯法**（conventional sawing 或 through and through sawing）

最快也最方便的下锯方法，原木沿着顺纹方向被平行锯开（图 7.2）。它有优点也有缺点。

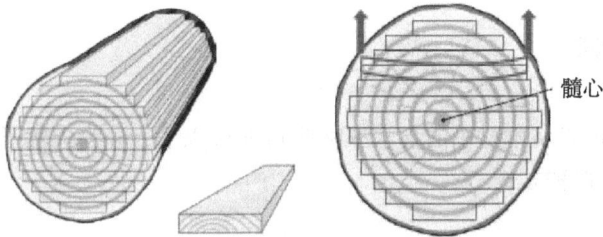

图 7.2 常规下锯法

常规下锯的优点为：成本低；废料少；加工板材方便，不用调整方向；加工便捷。

常规下锯的缺点为：板材干燥时容易杯弯；板材无特殊纹理；边材多，耐久性较差；不适合用作结构材。

（2）**四等分法**（quarter sawing）**或径切下锯法**（radial sawing）

原木先四等分，这种下锯方法可以获得银光纹理。这样下锯每锯一次，要翻转一次原木（图 7.3）。

四等分下锯或径切下锯的优点为：可以产生漂亮纹理；板材更稳定、干缩小；板面砂光更均匀。

四等分下锯或径切下锯的缺点为：劳动密集，因为要四等分并且不断翻转原木；成本高；废料多；难以产生宽料。

（3）**弦切下锯法**（Tangential sawing）

沿着原木的年轮切线方向锯切，以此方法锯切得到的锯材具有火焰状纹理或"V"字形纹理（图 7.4）。

弦切下锯的优点为：可以产生火焰状纹理；板材干燥更快；板材更好砂光；端部钉连接不易开裂。

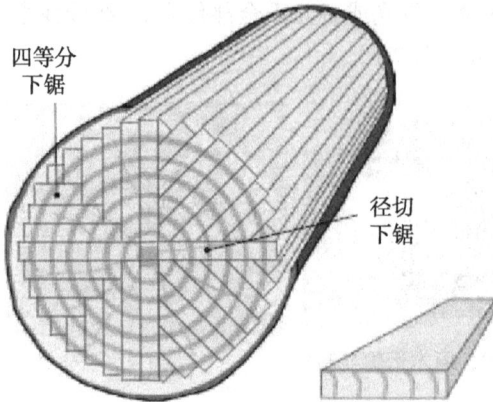

图 7.3　四等分下锯或径切下锯法

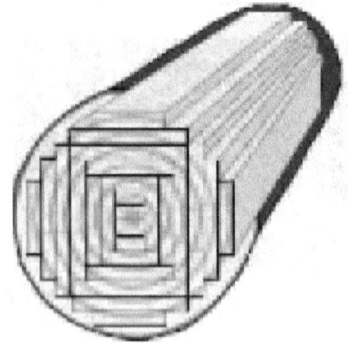

图 7.4　弦切下锯法

弦切下锯的缺点为：易干缩变形；木材易翘曲；加工成本高，因为锯切一次，反转 90°。

7.2　锯材干燥

木材干燥的目的是将木材中的含水率降低到 20% 以下，分为**天然干燥**（seasoning）和 **人工干燥**（drying）两种方式。

7.2.1　天然干燥

木材天然干燥（seasoning）是将木材堆放在空旷的板院内或通风的棚舍下，利用大气中的热力蒸发木材中的水分使之干燥（图 7.5）。

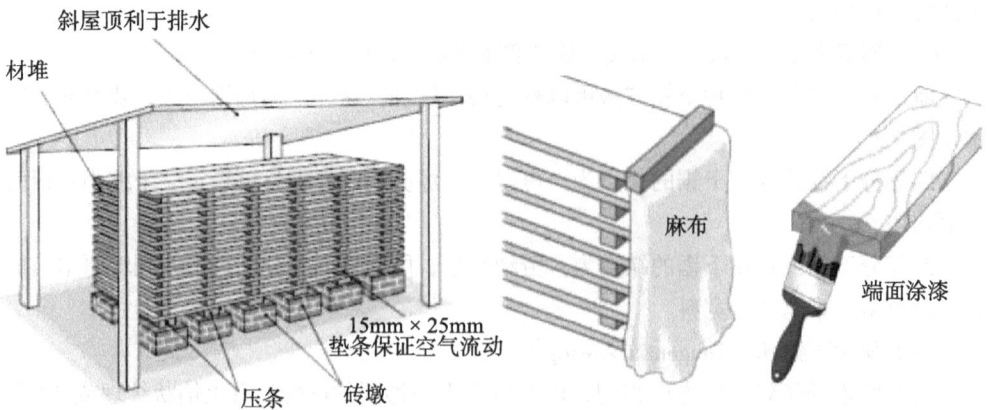

图 7.5　气干及板材端部处理

　　材堆（stack）用**垫条**（sticker）分开，以保证空气流通。垫条为横截面尺寸为 15mm×25mm 的木条，根据木材变形难易程度，进行排布。要注意的事项如下：

　　①在一个材堆内，最好堆放同一树种、同一厚度的木料，木料数量少时也可将材质相近的木材堆积在一起。

　　②木料应先分类，分别堆积，长材置于材堆外边，短材放在材堆里边，木堆的两端应堆齐，上下垂直。

　　③材堆在板院内应按主风方向来配置，即薄而易干的材堆放置在迎风的一边，中等厚度的材堆放置在背风的一边，木料厚而难干的材堆放置在板院的中部。

　　④材堆的长度应与主风方向平行。

　　⑤为使材堆中气流很好地循环，在堆积时木料之间要留空隙，上下对应，形成垂直气道。

　　⑥板的端面应避免日晒，防止端裂。

　　天然干燥仅能将板材含水率降至 18%~22%。具有这一含水率的木材可用于户外构件，若用于室内则需进一步人工干燥。

7.2.2　人工干燥

　　常用的人工干燥是窑干（kiln drying）。窑干是指在特制的建筑物或金属容器内，人为地控制干燥介质的温度、湿度及气流循环速度，主要利用气流介质的对流传热，对木材进行干燥处理，这是国内外广泛采用的一种干燥方法。

　　木材在**推车**（trolley）上像天然干燥那样堆积，再送入**干燥窑**（kiln）（图 7.6）。在墙壁、地板和顶棚管道上的**蒸汽喷射器**（steam jets）加热蒸汽，使之通过材堆。蒸汽仅加热木材而不干燥木材。一旦加热，窑内的相对湿度降低同时保持加热。这样就使得木材中的水分逐渐蒸发，直至达到最终含水率。风机可以促进窑内空气循环。气道用于湿空气的排出和新鲜空气的进入。

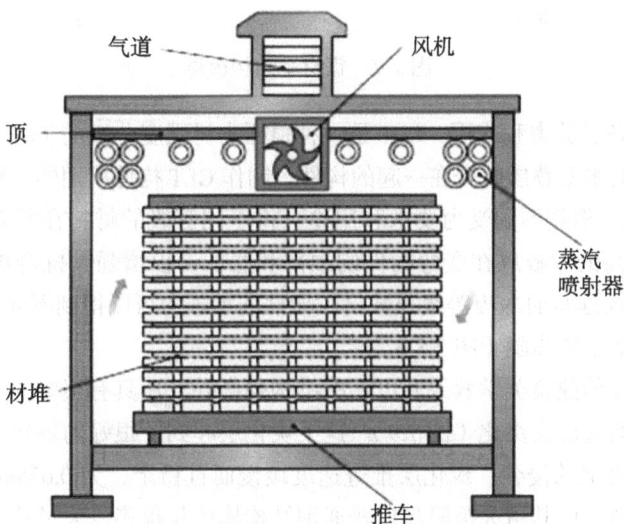

图 7.6　干燥窑

7.3 木质人造板加工

人造板材是以木材或其他植物纤维为原料经专门的工艺过程加工分离成各种形状的单元施加胶黏剂或不加胶黏剂，在一定条件下再组合压制而成的板材或型材。根据使用用途，可以分为结构人造板材与非结构人造板材。本教材重点介绍结构人造板材。

7.3.1 结构人造板材加工

（1）锯材基结构板材加工

锯材基结构板材（lumber-based engineered wood products）主要有：**层板胶合木**（glue-laminated timber，GLT）、**正交胶合木**（cross-laminated timber，CLT）、**层板钉合木**（nail-laminated timber，NLT）、**层板销合木**（dowel-laminated timber，DLT），如图 7.7 所示。在这些材料中，GLT **和** CLT 是两种常用的结构材料。

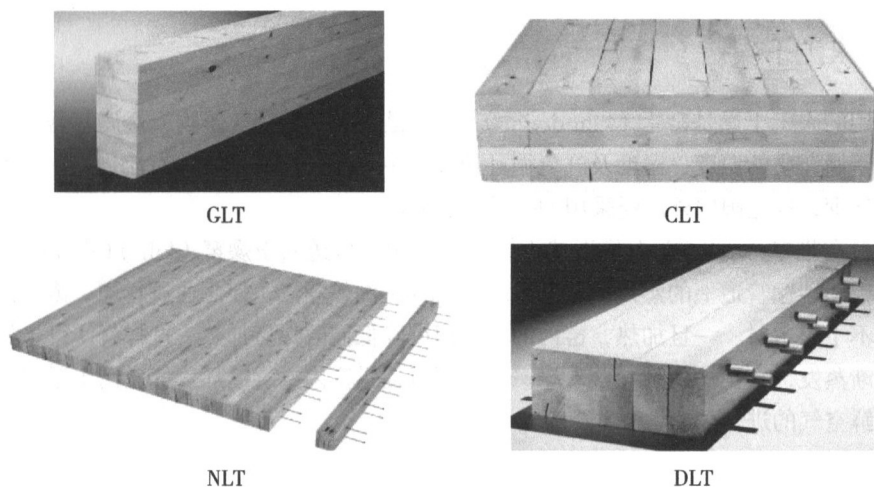

GLT   CLT

NLT   DLT

图 7.7 锯材基结构板材

① GLT 的制备、分类和应用　GLT 是一种根据木材强度分级的工程木产品，通常是由二层或二层以上的木板叠层胶合在一起的构件。制作 GLT 构件所用的木板，经过干燥和**分等分级**（grading），根据不同受力要求和用途，将不同等级的材料在截面方向进行同纹理组合。强度等级高的木板放在受弯构件的顶部和底部，以增强其抗弯承载力。根据构件在受拉区或受压区排布的木板等级及数量，构件的承载力可以得到不同程度的提高。GLT 的生产过程有以下主要步骤（图 7.8）。

GLT 除了本身的建筑美学效果和可靠的结构强度外，还具有很好的耐火性。火灾出现时，胶合木构件的外层会炭化（图 7.9）。这个炭化层起到了很好的隔热效果，保护了构件内部进一步受到火焰的侵袭。炭化层推进速度缓慢而且稳定，为 0.635mm/min。木结构构件虽然是可燃材料，但其耐火极限却比普通钢结构构件长很多。大尺寸的木结构构件在耐火方面时间更长。

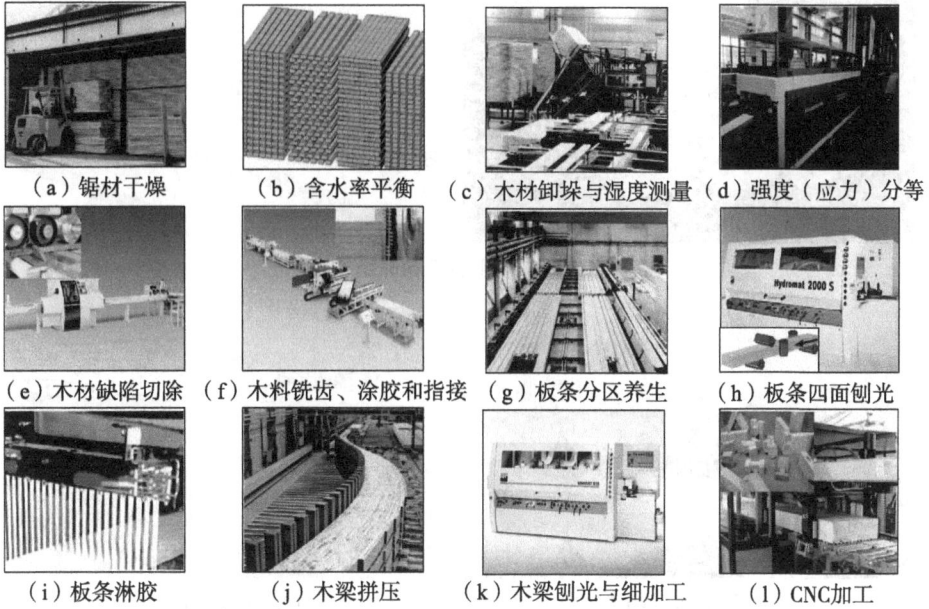

（a）锯材干燥　（b）含水率平衡　（c）木材卸垛与湿度测量　（d）强度（应力）分等

（e）木材缺陷切除　（f）木料铣齿、涂胶和指接　（g）板条分区养生　（h）板条四面刨光

（i）板条淋胶　（j）木梁拼压　（k）木梁刨光与细加工　（l）CNC加工

图 7.8　GLT 生产流程

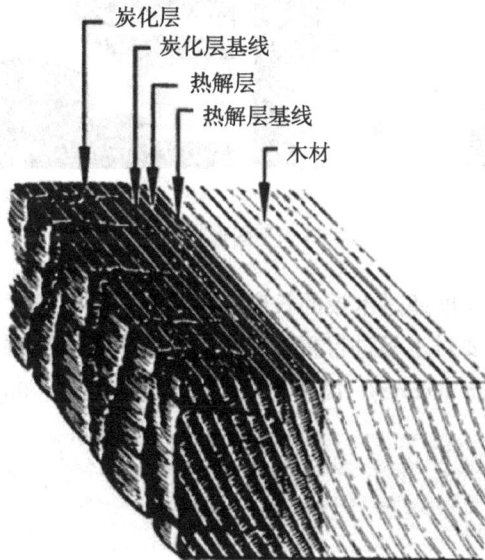

炭化层
炭化层基线
热解层
热解层基线
木材

图 7.9　木材炭化截面

GLT 根据承载情况分为：结构用 GLT 和非结构用 GLT（图 7.10）。结构用 GLT 主要用于承重领域，而非结构 GLT 用于家具、地板等非承重领域。

GLT 根据产品形状分为：圆形 GLT、弧形 GLT、方形 GLT、矩形 GLT 和异型 GLT（图 7.11）。

图 7.10　结构用 GLT 和非结构用 GLT

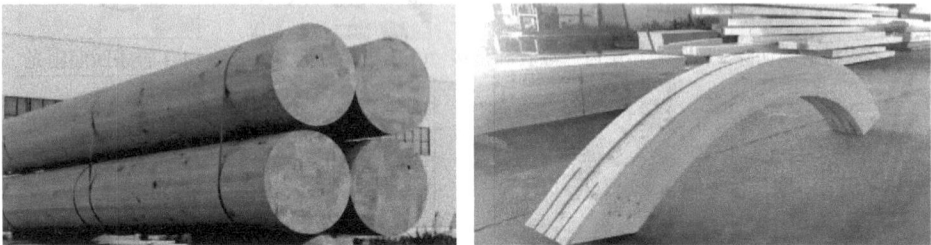

（a）圆形 GLT　　　　　　　　　　（b）弧形 GLT

（c）方形 GLT　　　　　　　　　　（d）异形 GLT

图 7.11　不同形状的 GLT

如果胶合木构件的长度超过木板的长度，则木板需要在长度方向进行端部拼接，或者为了去除缺陷，也需要对木板进行重新胶拼。根据木板端部拼接方式，可分为：**平接**（butt jointing）、**斜接**（scarf jointing）、**齿接**（tooth jointing）、**指接**（finger jointing）（图 7.12）。

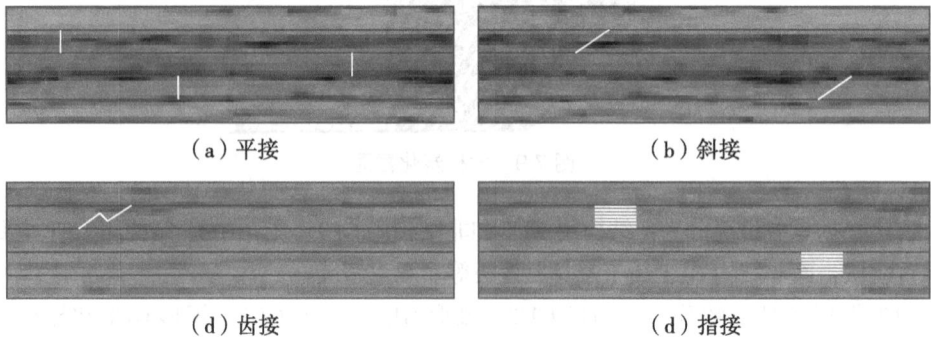

（a）平接　　　　　　　　　　　　（b）斜接

（d）齿接　　　　　　　　　　　　（d）指接

图 7.12　木板端部拼接方式

平接几乎没有任何受拉承载力，所以在受弯、受拉以及弯曲构件中不允许采用。对于木板的斜接，当接口坡度为 1/12 时，连接处的强度可以达到木板本身的强度。齿接可以满足快速生产的需要，但强度不如斜接。指接是近年来在结构用 GLT 中主要采用的木板端部拼接方式，但其强度不及斜接的强度高。

GLT 根据需要可以制成材质等级对称与非对称的构件，如图 7.13 所示。No.1、No.2、No.3 和 T.L 分别表示不同的层板等级。对于受弯构件，截面中最关键的是受弯方向的边缘受拉区。在不对称 GLT 梁中，受拉区边缘部位的木材材质等级比受压区木板材质等级高，这样可以有效提高木材的利用率。所以，不对称胶合梁的抗弯强度值对于受拉区和受压区是不同的。

图 7.13　对称结构 GLT 和非对称结构 GLT

GLT 没有改变木材的结构和特点，但力学强度和材料性能均匀性方面都优于实体木材。因此，GLT 可以代替实体木材，应用于实体木材所应用的各个领域（图 7.14）。

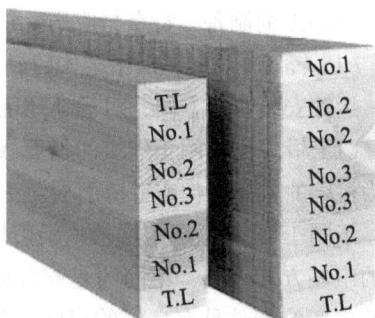

图 7.14　GLT 在结构和非结构领域的应用

②CLT 的制备、特点与应用　CLT 至少由 3 层实木锯材或结构复合材以一定角度组坯[图 7.15（a）、（b）], 采用结构胶黏剂胶合而成的预制工程木产品。CLT 具有预制化程度高、运输安装快捷方便，对场地环境影响小的特点，被认为是传统建筑材料的最佳替代产品。CLT 板广泛应用于建筑的**墙面板**（wall）、**楼面板**（floor）和**屋面板**（roof），同时还可作为桥梁的主体结构或者桥面板。

不同于 GLT 每层的层板均沿着顺纹组坯，CLT 是采用纵横交错的组坯方式，和胶合板的组坯方式一样，不同的是胶合板用的是单板，而 CLT 用的是锯材。用这样的组坯方式所生产的产品具有良好的双向力学性能以及卓越的尺寸稳定性。此外为满足工程上的特殊需要，可以沿相同纹理方向放置双层木板，从而在顺纹方向具有优越的力学性能。为满足更加多样化、复杂的结构形式，加拿大学者提出盒式 CLT 板 [图 7.15（c）], 该体系在保证具有良好承载能力的基础上能有效降低 CLT 板重量，并具有良好的工程应用价值。

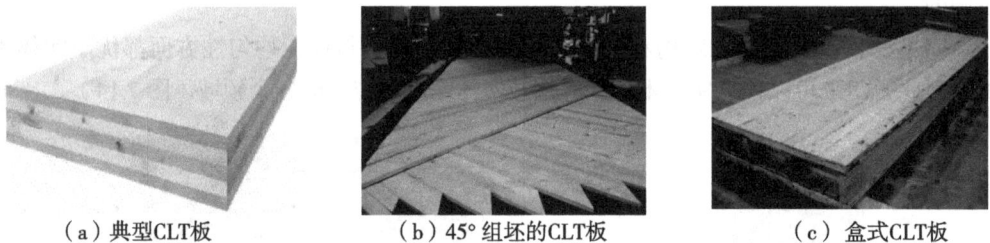

（a）典型CLT板　　　　（b）45° 组坯的CLT板　　　　（c）盒式CLT板

图 7.15　CLT 结构形式

CLT 成品板材按照生产流程可分为 11 个步骤（图 7.16），其中对成品板材各项性能指标影响较大的工艺主要有以下 3 个方面：原料分选加工、施胶组坯及成型加压。

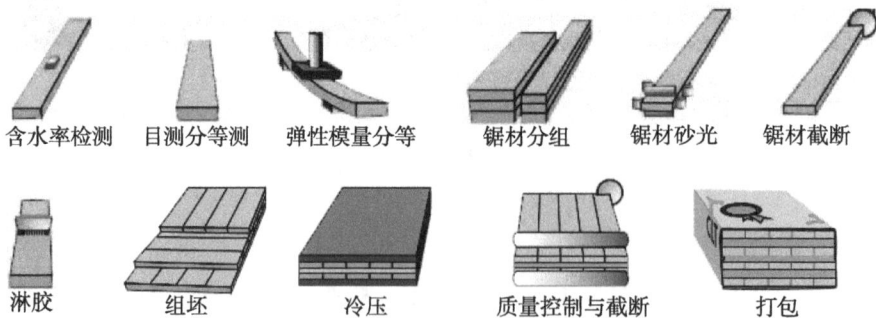

含水率检测　　目测分等测　　弹性模量分等　　锯材分组　　锯材砂光　　锯材截断

淋胶　　　组坯　　　　冷压　　　质量控制与截断　　　打包

图 7.16　正交胶合木生产流程

原材料的含水率和材料本身等级（按节子、虫蛀、腐朽等缺陷相对于原材料本身所占比例进行划分）是决定 CLT 性能的重要影响因素，北美 ANSI/APAPRG 320 胶合木规范对普通胶合木层板材质有明确的等级标准划分，主要包括针对不同材料的弯曲强度、剪切强度等力学要求以及表面质量分等方面。

图 7.17（a）为加拿大英属哥伦比亚大学内的 18 层学生公寓 Brock Commons，采用

的是框架－核心筒体系，核心筒为钢筋混凝土现浇而成，第 1 层采用混凝土框架结构，第 2~18 层采用 CLT 作为水平受力构件，层间采用 GLT 柱作为竖向受力构件，板构件与柱构件之间采用钢构件进行连接，建筑高度达到 54m。世界高层建筑与都市人居学会（CTBUH）宣布，2019 年 3 月完工的挪威 Mjøstårnet 是世界最高木结构建筑[图 7.17（b）]。该建筑高度达到 85.4m，共有 18 层，同时为避免在风荷载作用下水平位移过大的问题，上半部分的楼板采用预制混凝土楼板。

（a）加拿大 Brock Commons　　　　　　　　　（b）挪威 Mjøstårnet

图 7.17　CLT 高层建筑

由于我国规范的不完备以及大众对于多、高层木结构建筑认知缺乏，现阶段 CLT 建筑在我国以低层为主，并且所建建筑以展示推广为目的。2014 年 3 月，国内首栋双层 CLT 和轻型木结构混合示范建筑在河北省迁安市落成，如图 7.18 所示。同年，位于我国台湾省的森科 5 层 CLT 建筑竣工。建筑师充分利用 CLT 良好的悬挑性能，在 2~4 层设置悬挑阳台，整栋建筑造型独特、美观，如图 7.19 所示。同时在室内将 CLT 暴露，体现出 CLT 本身特色，为亚洲首栋多层 CLT 建筑。

图 7.18　国内首栋 CLT 和轻型木结构　　　　　图 7.19　台湾森科大厦
混合示范建筑

图 7.20 为宁波中加低碳新技术研究院有限公司与同济大学合作建造的中国第一栋 CLT 示范公共建筑—OTTO Café。此外，宁波中加低碳新技术研究院有限公司携手葡萄牙阿默林集团，应用生态软木板（insulated cork board，ICB）作为外墙的隔热节能和装饰材料旨在提升 CLT 民居（度假屋）的绿色品质（图 7.21）。CLT 建筑在中国旅游地产、城市化建设和农房工业化等领域的示范和推广有着巨大优势。

图 7.20　OTTO Café

图 7.21　CLT 民居

③GLT 和 CLT 的异同　GLT 是一种传统的工程木材，采用规格锯材胶合而成。GLT 中所有层板锯材皆以顺纹组坯，可用作直或弯或带一定弧度的梁柱构件。CLT 是近几年风靡全球的新型预制建材，是锯材层板纹理正交组坯胶合而成的结构板材。

从原材料来看，用于 GLT 和 CLT 的树种可以是相同的，主要是针叶树种，包括：挪威云杉、白冷杉、欧洲赤松、欧洲落叶松、花旗松、西部落叶松。云杉—松—冷杉（SPF）和挪威云杉分别是北美和欧洲常用的胶合木原料。当然，也有一些阔叶树种用作胶合木的生产。所使用的结构胶黏剂主要有 EPI、PUR、PRF。

从生产工艺来看，CLT 板可以在大幅面压机中进行四面加压成型，周期为 1 小时左右（PUR 胶）；而 GLT 的生产需要侧向施压夹持养护，需要一周以上养护时间。从应用角度来看，CLT 主要用于楼板、墙板、屋面板，较少用于梁柱构件；而 GLT 主要用于梁柱构件。

④层板钉合木　层板钉合木（NLT）是将规格材侧面拼接，并用钉子或螺钉紧固，制成尺寸较大的平板结构材。与胶合木（GLT）相同，NLT 也是单方向拼接（图 7.22）。在北美 NLT 和其他层合板类似，每层的厚度为 38mm 以上、宽度为 64mm 以上。NLT 制造时的含水率通常为 12%~16%，选用目测或机械分等（如应力分等 MSR）的软木规格材，如 2 级规格材或 1650f~1.5E 级 MSR 材。如果预制板材的长度在 6m 以内，只需用拼板即可；如果超过 6m，就需要用胶接或指接木材。所以，在制造 NLT 时，不用或极少量使用胶黏剂。

（a）线性连接

（b）错位连接

图 7.22　层板钉合木的连接方式

　　NLT 的优点在于适合现场组装；在运输或持续的动态力作用下（如车轮载荷和往复运动的工业设备），钉子能够吸收能量阻尼振动；NLT 还具有很好的防火性能。NLT 的缺点是，当施加较高的刚性载荷或作为弯曲构件时，机械制造效率会变差；拼板间钉子弯曲以及连接容易产生缝隙（各层间的水分转移导致的裂缝）；组装速度慢；另外，因金属钉子的存在，NLT 组装后难以切割。

　　作为一种重型木基结构材，近年来 NLT 的使用率有所提高。在一些复杂的木结构建筑的设计和建造中，可以选用 NLT。图 7.23 所示的是一家位于中国青岛的旅游接待中心，其屋顶是以 NLT 为主体，结合规格材、胶合板和 OSB 等覆面板来加固完成整体结构。覆面板用钉子紧固于 NLT 上，这种隔板结构可以抵抗侧向力，同时使整个系统处于干燥状态。图 7.24 所示的是美国明尼阿波利斯 T3 木结构大楼，高 7 层，其中楼板使用了 NLT，梁合柱选用了 GLT。

图 7.23　中国青岛旅游接待中心 NLT 屋顶

图 7.24　美国明尼阿波利斯 T3 木结构大楼，其中楼板用 NLT 制成

　　⑤层板销合木（DLT）　DLT 和 NLT 的拼板相同，但连接的方式不同。DLT 用木销紧固。DLT 也是单方向拼接。现代 DLT 是 20 世纪 90 年代由瑞典人首先发明的。DLT 使用厚 38mm，深度为 89mm、140mm 或者 184mm 的软木规格材，通过侧面拼接，面与面间用木销连接。与 NLT 不同的是，拼接 DLT 主要用指接材，其拼接时的含水率在 19% 以下。

　　木销通常为高密度的阔叶树材（如栎木），直径一般为 19mm，含水率约为 6%~8%。将木销插入拼板之前，要做出直径相同的预留孔。木销可以线性排列或交位排列，如图 7.25 所示。销间间隔为 300mm，错位排列的 DLT 刚度更高。

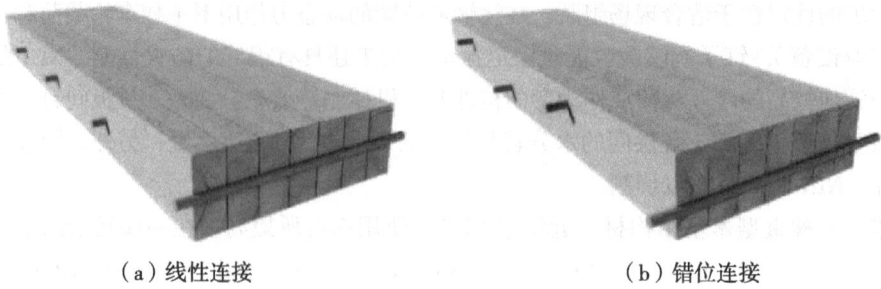

（a）线性连接　　　　　　　　　　　　（b）错位连接

图 7.25　层板销合木的连接

DLT 板材上可加工线管和其他孔，实现个性设计，如图 7.26 所示。设计者通过加工切口和凹槽来提高声学性能和美化外观。例如，吸声条可以安装到 DLT 板材的内部，这样可以降低噪音，又不影响外部纹理形态，同时还可以进行其他外部装饰。

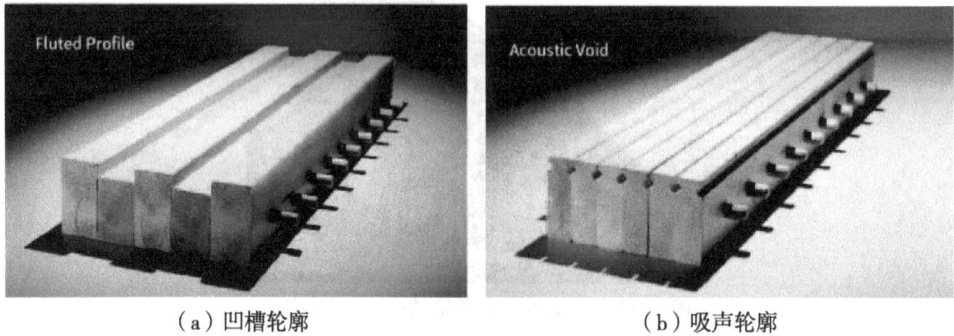

（a）凹槽轮廓　　　　　　　　　　　　（b）吸声轮廓

图 7.26　木销连接层积材的异形轮廓

DLT 很适合用于地板和屋顶建造，建造中的七层高美国亚特兰大 T3 木结构大楼，其中楼板合和屋顶均用 DLT 制成，如图 7.27 所示，同时也可以用于墙体。如果在 DLT 的上部增加多层胶合板，就可以获得双层结构。另外，DLT 可以用于承重构件和剪切墙，电梯和楼梯踏板的构建。DLT 的设计要求可以参照 NLT，目前尚无可参考的产品规范。

图 7.27　建造中的美国亚特兰大 T3 木结构大楼

（2）单板基结构板材加工

①结构胶合板和单板层积材加工　结构胶合板（structural plywood）和单板层积材（laminated Veneer lumber, LVL）的结构上的区别在于组坯结构不同，结构胶合板是正交组坯，而单板层积材为顺纹组坯（图 7.28）。木材原料主要有：花旗松、落叶松、南方松、黄杨、西部铁杉、美国黑松和云杉等，胶黏剂主要为酚醛树脂等。

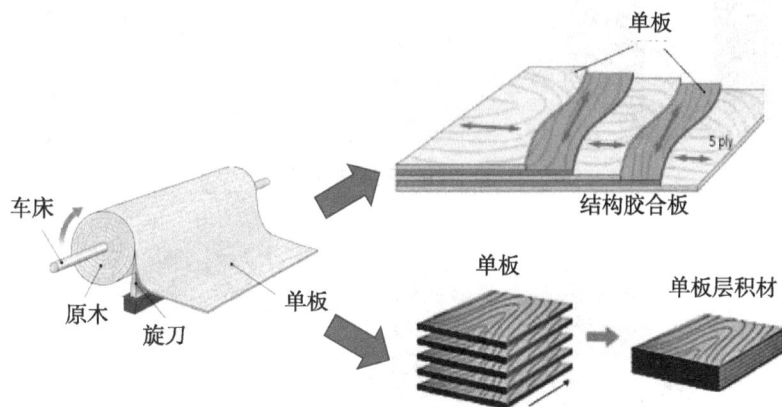

图 7.28　单板基结构板材

和结构胶合板生产工艺流程类似，单板层积材的生产工艺流程包括原木蒸煮、单板旋切、单板干燥、单板分等、涂胶、热压、修剪、切割、成品分等、打包运输等环节。

在生产开始之前，需要对原木进行剥皮处理，再截成一定尺寸的木段。再对木段进行水热或蒸汽处理，使木材软化利于单板制备，大部分针叶树材木段芯部温度要达到 50~60℃，单板的加工可以使用旋切和刨切这 2 种方式。结构胶合板和单板层积材使用的单板一般采用旋切加工，而非结构的装饰人造板多用刨切加工。单板厚度控制在 2.5~4.2mm。

现代化的胶合板或单板层积材工厂采用的是高速旋切机，单板带很长，需要剪裁成适宜的宽度，一般为 1.37m。有缺陷的单板或者尺寸不够的单板会被自动挑出，可用于其他木质材料的制备。单板随后进入干燥机进行干燥，最终到达 3%~6% 的含水率。

干燥后的单板要进行超声波应力分等，最高等级用于制备单板层积材，次等单板用于胶合板。单板一般使用涂胶或喷胶方式进行施胶后再进行组坯。大部分胶合板工厂在热压工序前都会进行冷压，以减少板坯厚度和单板错动。

②单板条层积材　原木旋切过程中产生的不完整单板，以及单板剪裁的剩余物，都可以制成木片，再施加酚醛树脂胶黏剂，然后在长度方向上定向组坯、热压形成新型结构板材 – 单板条层积材（parallel strand lumber, PSL）（图 7.29）。生产 PSL 的木片尺寸一般为宽度 20mm、厚度 3~4mm、长度 1~2.44m。PSL 主要用作木结构的梁和柱。

（3）木片基结构板材

定向刨花板（oriented strand board, OSB）（图 7.30），定向刨花板看起来和刨花板类似，但具有胶合板的特点。将原木加工成木片，然后热压胶合而成。表层木片和芯层木片长度方向相垂直，类似于三层胶合板结构。主要用于产品包装、地板、家具和木结构房屋中墙体和楼板的覆板。

图 7.29　PSL 及应用

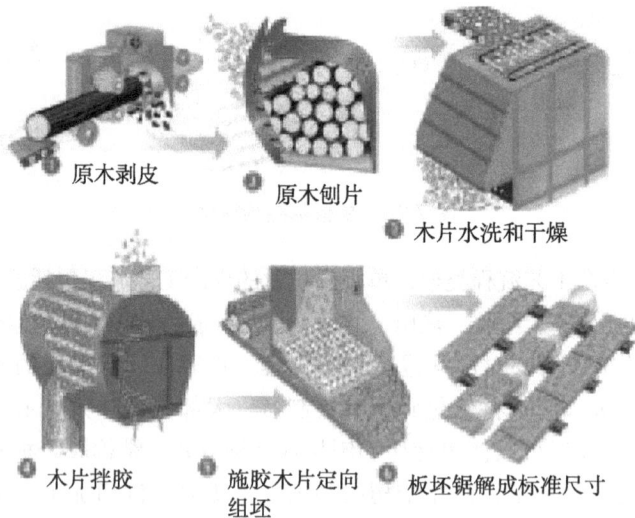

原木剥皮　　　原木刨片　　　木片水洗和干燥

木片拌胶　　　施胶木片定向　　板坯锯解成标准尺寸
　　　　　　　组坯

图 7.30　OSB

　　层叠木片层积材（laminated strand lumber, LSL）[图 7.31（a）]，和 PSL 类似，不同在于它使用的是更薄、更宽的木片和胶黏剂。使用改进的刨片机将白杨、黄杨和椴木或者其他低密度的阔叶树材加工成木片。木片的尺寸长度大约 30cm，是 OSB 用木片的两倍。一般使用异氰酸酯（MDI）胶黏剂。采用蒸汽喷射热压。由于异氰酸酯与水发生反应，因此木片可以保持较高的含水率（15%）。LSL 板经过切割后，主要用作梁、过梁（lintel）及墙骨柱（stud）。

　　定向木片层积材（oriented strand lumber, OSL）[图 7.31（b）]，是采用类似 OSB 的生产工艺制成的工程木制品。不同的是，OSL 中木片是在一个方向定向。而和 LSL 的区别在于，OSL 使用的木片更短些，长度只有 15cm，这种木片尺寸和 OSB 使用的木片是一样的，但更规整。OSL 的用途和 LSL 相似，主要用作梁和过梁。

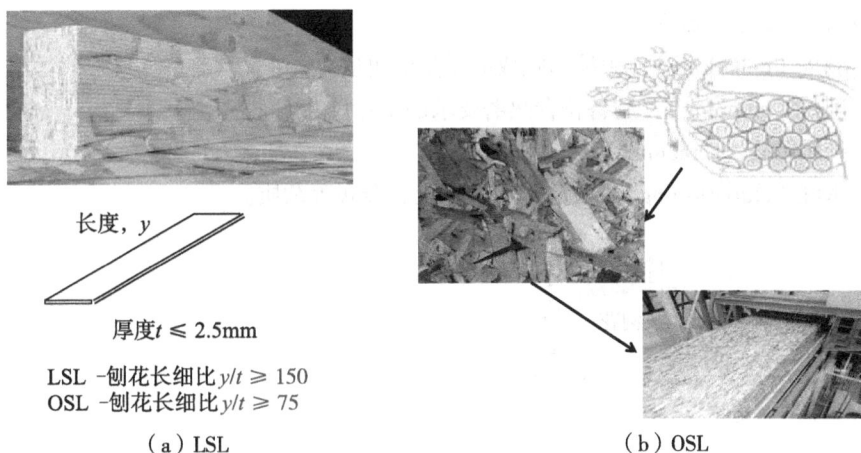

长度，y

厚度 $t \leqslant 2.5mm$

LSL -刨花长细比 $y/t \geqslant 150$
OSL -刨花长细比 $y/t \geqslant 75$

（a）LSL

（b）OSL

图 7.31 LSL 和 OSL

7.3.2 非结构人造板材加工

（1）细木工板加工

细木工板是由木条或木块顺纹方向组成板坯，上下面或侧面与单板或实木板组坯胶合而成的一种人造板材（图 7.32）。

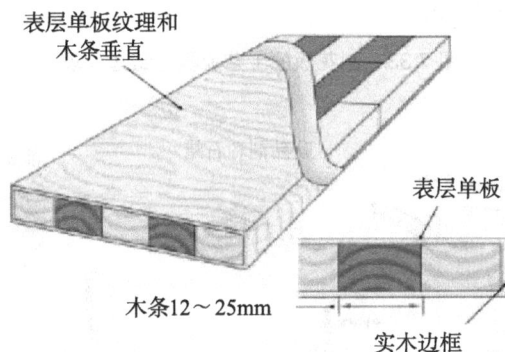

表层单板纹理和
木条垂直

表层单板

木条12～25mm

实木边框

图 7.32 细木工板

（2）刨花板加工

刨花板也叫颗粒板、碎料板（图 7.33），将各种小径木、速生木材、枝桠材等切削成一定规格的碎片，经过干燥，拌以胶料、硬化剂、防水剂等，在一定的温度压力下压制成的一种人造板，刨花或碎料排列不均匀。刨花板一般要贴面，可以外贴装饰薄木，用于家具等。

（3）中密度纤维板

中密度纤维板（medium-density fiberboard，MDF）是一种常用的人造板材。它具有标准的幅面尺寸和厚度尺寸。是将木材或植物纤维经机械分离和化学处理手段，掺入胶黏剂和防水剂等，再经高温、高压成型制成的一种人造板材，是制作家具较为理想的人造板材（图 7.34、图 7.35）。

MDF 的主要工艺流程如下：

① 剥皮（debarking）：去掉影响成品质量的树皮。

② 削片（chipping）：制备符合热磨要求的木片尺寸。

③ 纤维筛查（screening）：去除过大木片。

④ 水洗（chip washing）：去除砂石、泥土、金属等杂质。

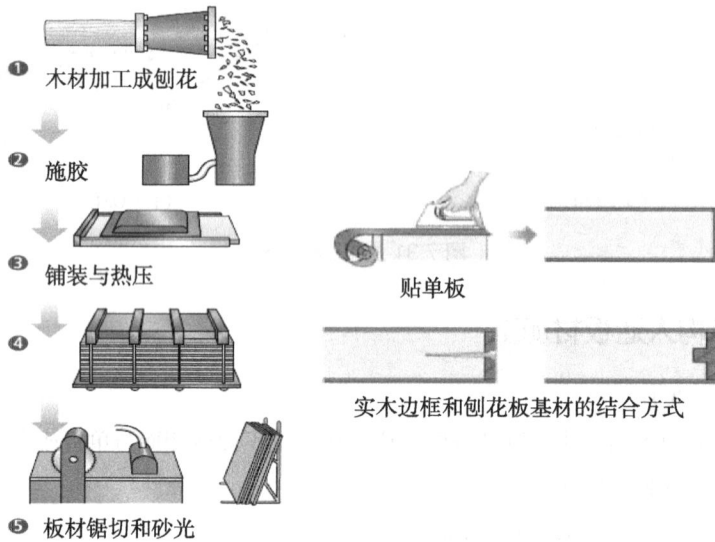

❶ 木材加工成刨花

❷ 施胶

❸ 铺装与热压

贴单板

❹

实木边框和刨花板基材的结合方式

❺ 板材锯切和砂光

图 7.33 刨花板的工艺流程与封边处理

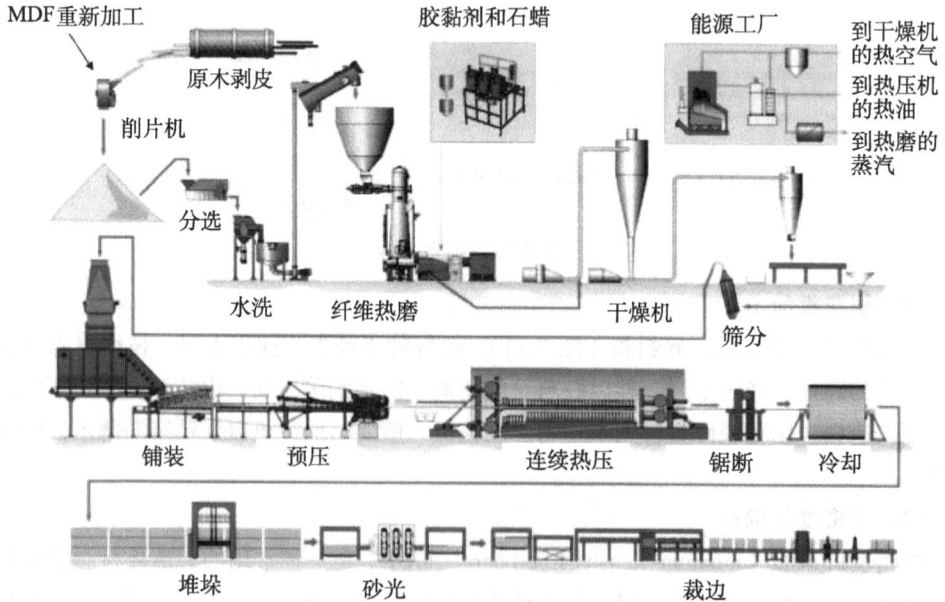

MDF重新加工

原木剥皮

削片机

分选

水洗

纤维热磨

胶黏剂和石蜡

干燥机

能源工厂

到干燥机的热空气
到热压机的热油
到热磨的蒸汽

筛分

铺装 预压 连续热压 锯断 冷却

堆垛 砂光 裁边

图 7.34 MDF 生产流程

⑤ 磨浆（defibration）：软化木片在磨盘的摩擦、挤压、揉搓等作用下，分离成单体纤维或纤维束。

⑥ 施胶（glue spreading）：对分离后的纤维，添加各种化学药剂，如胶黏剂、防水剂等，确保纤维结合，提高纤维板物理力学性能和耐水性。

⑦ 纤维干燥（drying）：纤维在"闪电式"管道干燥机（"flash" tube drier）中含水率降至 5% 以下。通过旋风分离（cyclone separation）将高速气流和木纤维分开。

图 7.35　MDF 板材

⑧ 纤维筛分（fiber sifter）：采用分选器，根据粗细纤维的重量不同，在一定的涡流运动中，产生不同的运动状态而达到相互分离的目的。

⑨ 板坯铺装（mat forming）：采用机械成型头或气流成型头对板坯进行铺装。

⑩ 预压（pre-compressing）：使板坯具有一定的密实度，提高初结合强度；可适当提高热压速度；减少板坯厚度。

⑪ 连续热压（continuous hot-pressing）：施胶后的板坯在一定的温度、压力和时间内成型的过程。

⑫ 后期加工：锯断、冷却、存储、砂光及裁成一定尺寸。

➤ 思考与训练

1. 原木制备成规格锯材需要哪些步骤？

2. 原木锯解有哪几种下锯方式？试比较优劣。

3. 木材干燥的方法有哪些？各自有什么特点？

4. CLT 和 GLT 有何异同？主要用途是什么？

5. OSB 和 OSL 有何异同？主要用途是什么？

6. LSL 和 LVL 有何异同？主要用途是什么？

7. 非结构人造板材有哪些产品？主要用途是什么？

第8章
木材缺陷

▶ **学习目标**

（1）了解木材天然缺陷；

（2）了解木材加工缺陷。

▶ **本章描述**

木材作为生物质材料，有哪些天然缺陷？如何辩证看待？木材加工过程中会产生哪些缺陷？本章将解答这些疑问。

木材是一种天然可再生的生物质材料，具有易加工、强重比高、热绝缘和电绝缘特性、纹理色调丰富美观、可循环利用等优点，被广泛用作木结构建筑材料、家具和室内装饰材料以及各类人造板的原材料。然而，生产木材的树木在生长过程中会因生长应力或自然损伤而形成节子（knot）和斜纹理（spiral grain）等天然缺陷；在加工木材的过程中会因干燥和机械加工引起干裂、翘曲、锯口等加工缺陷；在利用木材过程中极易受虫菌侵蚀而引起腐朽、变色和虫蛀等缺陷。凡是呈现在木材上能降低其质量、影响其使用的各种缺点，都可以称为**木材缺陷**（wood defects）。因此，认识和了解木材缺陷及其特性，对木材进行改性与保护，如防腐、阻燃、强化处理，是提高木材质量，延长木材使用寿命，提高木材利用水平，节约木材资源的重要途径和手段。

需要注意的是，对于木材缺陷应该辩证看待，如木材节子与应力木等自然特性会降低木材力学性能，但赋予了木材多变纹理，从而有助于形成美丽的木纹，提升木材的装饰效果；木材的生物降解性尽管使木材耐久性降低，但木材生物降解有助于木材回归自然，因而具有低碳环保之性能。总而言之，木材利用应在尊重木材自然特性的前提下，做到"因用改材"或"因材适用"。

8.1 木材天然缺陷

木材的天然缺陷是由于树木生长的生理过程、遗传因子的作用或者在生长期受到外界环境的影响而形成，主要包括节子、应力木、裂纹、树干形状缺陷等。

8.1.1 节子

包埋在树干或主枝木材中的枝条部分称为**节子**（图 8.1）。树木从一棵幼苗长成大树，不断地从髓心生出小枝，随树干逐渐加粗，枝条被包藏起来，形成树节。节子是木材中存在的一种天然生长特性，因其影响木材的强度等，通常被认为是一种缺陷。但节子及其周围的纹理，又可以在装饰中起到美化作用。

节子的数量和大小是决定木材等级的最主要因素。按照节子与周围木材连生程度来分，节子可以分为：活节下页①和死节下页②（图 8.2）。活节是活树枝所形成的节子，活节的木材组织与其周围木材组织全部紧密相连，质地坚硬，构造正常。而死节是死树枝所形成的节子，节子的木材组织与其周围木材组织部分或完全脱离。

若按节子材质分类，可分为：健全节、腐朽节和漏节。

健全节：节子没有腐朽，颜色与周围木材颜色一致或稍深。

腐朽节：节子本身已经腐朽，但未透入树干内部，节子周围材质仍完好。

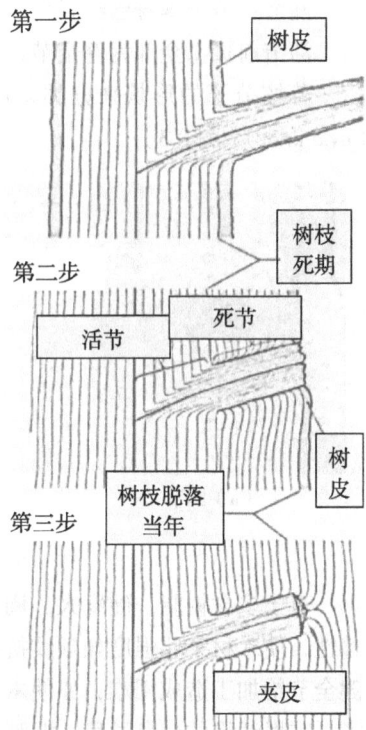

图 8.1 节子的形成

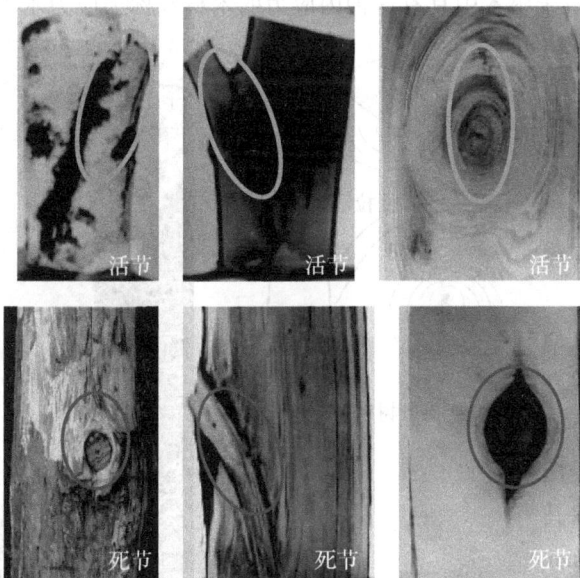

图 8.2 活节与死节

漏节：节子本身已经腐朽，而且深入树干内部，引起木材内部腐朽。

活节前页①全部为健全节，而死节前页②可能是腐朽节或漏节。

若按节子的形状来分类，可分为：圆节（round knot）、尖节（spike knot）和椭圆节（oval knot）（图 8.3）。

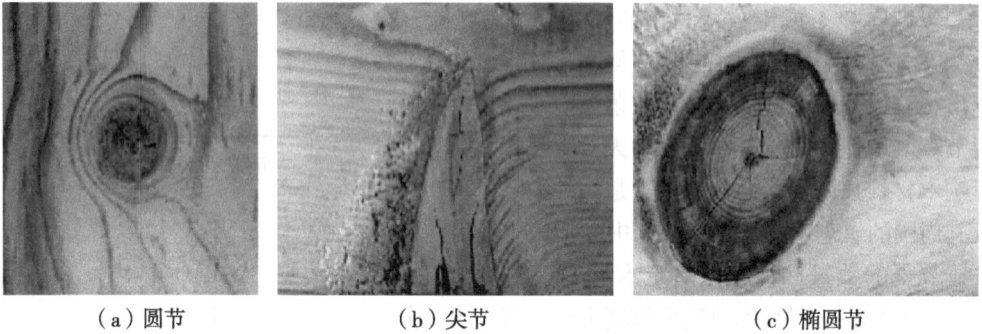

（a）圆节　　　　　　　　　（b）尖节　　　　　　　　　（c）椭圆节

图 8.3　节子的形状

节子多含树脂，硬度大，周围木材纹理局部紊乱。节子破坏了木材结构的均匀性及完整性，使木材某些强度如顺纹抗拉、抗弯强度降低，不利于木材的有效利用。活节与死的健全节给加工造成困难，如使木材纹理紊乱、增大刀具的切削阻力、制浆造纸时节子难熬煮、减慢纤维的分离过程、混脏木浆、影响纸张颜色等。

8.1.2　应力木

树木为了保持树干笔直生长、或者使树枝恢复到正常位置，会在树木的某些部位产生一种内应力。来自于树木中具有这种内应力的木材被称为应力木（reaction wood）（图8.4）。它在解剖构造和材性上与正常材有显著的差异，这部分木材横截面上的生长轮通常特别加宽，而其相对的一侧生长轮则表现正常或狭窄。

在针叶树中因偏宽年轮位于倾斜树干及树枝的下侧，这部分木材组织在立木时期受压应力作用，因而称其为应压木（compression wood）。相反，阔叶树中的偏宽年轮位于倾斜

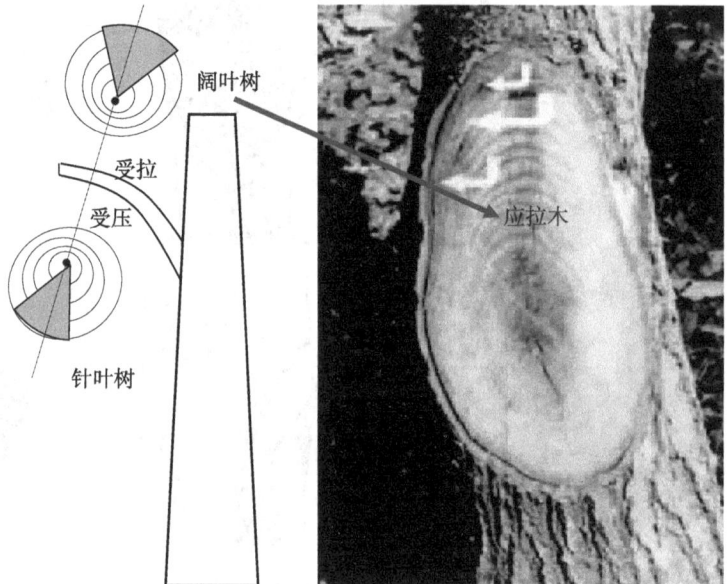

图 8.4　针叶树和阔叶树应力木的差异

树干及树枝的上侧，受的是拉应力作用，则称其为应拉木（tension wood）。表 8.1 总结了应压木和应拉木的主要特征及其加工中常出现的问题。

表 8.1 应压木和应拉木

应力木	主要特征	主要问题
应压木	管胞胞间隙增大 S_2 层纤丝角增大	板材易发生严重扭曲或翘曲 木材顺拉强度、静曲强度下降
应拉木	出现胶质木纤维 木质素含量下降	锯剖和旋切时板面起毛 锯切时易夹锯 木材强度下降

8.1.3 裂纹

树木在生长时期或伐倒后，因受外力[①]作用或温度、湿度变化的影响，木材纤维与纤维之间发生分离所形成的裂隙，叫裂纹（check）或开裂（split）。按照类型和特点分为径裂和轮裂。

◎ 径裂或心裂（图 8.5） 沿半径方向开裂的裂纹。径裂又可分为单径裂（一条或两条裂纹在同一直径上）和复径裂；复径裂也称为辐射状径裂或星裂（几条裂纹从髓心向各个方向辐射），此外还有在极端气候条件下的冻裂。

◎ 轮裂（图 8.6） 木材的轮裂是在木材断面沿年轮方向开裂的裂纹。轮裂又分为弧裂（指开裂占年轮一部分）和环裂（指沿年轮开裂）两种。

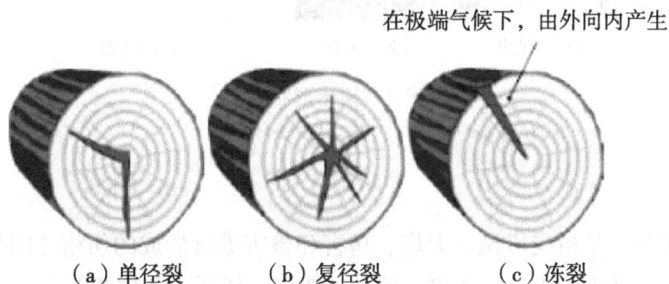

图 8.5 径裂

（a）单径裂　（b）复径裂　（c）冻裂

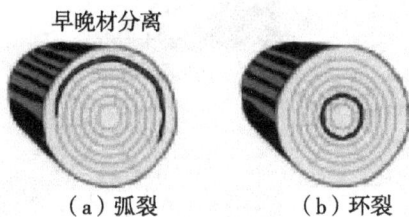

图 8.6 轮裂

（a）弧裂　（b）环裂

---

① 两种外力，一种是生长应力，另外一种是水分应力。

在木材中，裂纹是指木材细胞间的分离，但没有贯穿其厚度；开裂是指木材细胞间的分离贯穿其厚度。裂纹在不适当的木材干燥和使用中，会扩展成为开裂。另外，就抗弯强度而言，轮裂的影响大于径裂。

8.1.4　树干形状缺陷

树木在生长过程中受到环境条件的影响，使树干形成不正常的形状，叫树干形状缺陷（图 8.7），有弯曲、尖削（taper）、树瘤（burr 或 burl）等。

◎ 弯曲：树干的轴线（纵中心线）不在一条直线上。

◎ 尖削：树干或原木直径大、头部直径小的直径相差比较悬殊的一种木材缺陷。

◎ 树瘤：指树干上局部木材组织增长不正常所形成的瘤状物。

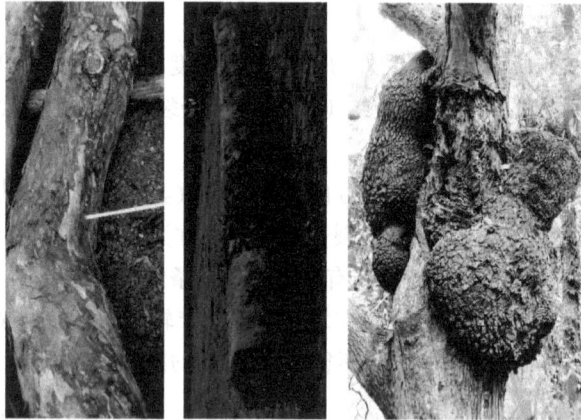

（a）弯曲　　　（b）尖削　　　（c）树瘤

图 8.7　树干形状缺陷

8.1.5　伤疤

伤疤也称损伤，是指受机械、火烧、鸟害和兽害等所形成的伤痕（图 8.8），主要包括机械损伤、烧伤、鸟害和兽害、夹皮、偏枯、树包、风折木和树脂漏等。

图 8.8　伤疤

8.2　生物危害缺陷

8.2.1　木材生物降解（biodegradation）

木材作为天然有机物，易受微生物的侵害，使木材发霉、变色、腐朽、分解，导致木材逐渐败坏而失去用途。微生物主要分为真菌和裂殖菌，而对木材败坏最严重的是真菌。寄生在木材上的真菌可以分为 4 类：木腐菌（decay fungus）、软腐菌（soft rot fungus）、木材变色菌（staining fungus）和霉菌（mould）。前两种真菌出于它们自身的生存和繁殖的需要而分解木材细胞壁，引起木材腐朽破坏。而后两种主要摄取细胞腔中的物质作为养分，因此对细胞壁无破坏。

真菌的生活周期由生长期和结实期组成，对材质的破坏是在生长期。在适宜的条件下，木腐菌的菌丝端头可以分解酶，将木材细胞壁溶解成小孔眼，于是菌丝便通过小孔而扩展。而变色菌的菌丝则主要是借细胞壁上的纹孔进入细胞腔，故对木材损伤不大。

真菌的生长必需具备如下条件，缺少其中任何一个条件，菌丝就不能生长或生长被抑制：

◎　适宜的温度（0~32℃）；

◎　有一定的氧气供给；

◎　足够的水分（含水率接近纤维饱和点）；

◎　适合的营养供给（也就是木材本身）；

防止木材腐朽和变色就是通过控制上述条件而实现的。

8.2.2　木材腐朽

一般来说，木材腐朽主要分为 3 类：褐腐（brown decay）、白腐（white decay）和软腐（soft decay），分别由褐腐菌（brown rot fungus）、白腐菌（white rot fungus）和软腐菌（soft rot fungus）引起。褐腐菌分解和破坏纤维素和半纤维素；白腐菌主要破坏木质素，少量破坏纤维素和半纤维素；软腐菌主要分解纤维素。

经褐腐菌侵害后，木材颜色变为褐色 [图 8.9（a）]，木材的重量将会大幅下降，纤维素含量将减少 90% 以上。腐朽后期，腐朽材成为深浅不同的褐色斑块，很容易把它捻成粉末，故又称粉末状腐朽。当木材重量损失超过 20% 时，木材的力学强度显著下降，且木材在径向和弦向就会发生明显收缩。

白腐菌主要分解木材中的木质素，这是由于白腐菌产生的酚氧化酶的氧化分解作用而破坏，纤维素是因真菌分泌的酸和酸水解而受到破坏。受害木材的材色为白或黄白，有的材质松软，状如海绵，称海绵状腐朽；有的显露出纤维状结构，犹如蜂窝状，称筛孔状腐朽 [图 8.9（b）]。

长期泡在不流动的水中，木材会受到软腐菌的破坏。软腐菌的菌丝侵入细胞壁中沿着纹理方向蔓延，且主要溶蚀木质化程度较低的次生壁 S_2 层，呈无数空洞（图 8.10）。软腐早期不借助显微镜则难于判定，软腐后期，木材表面通常褪色，材面变软，刮脱后露出下层相对健全的木材。

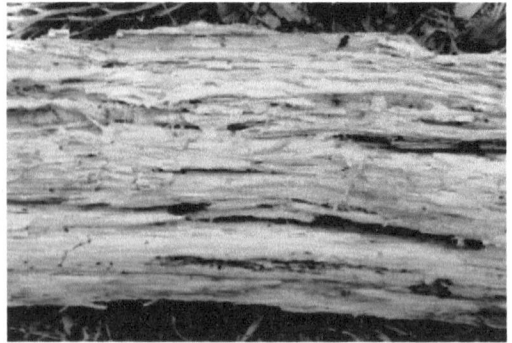

（a）褐腐　　　　　　　　　　　　　　　（b）白腐

图 8.9　木材腐朽

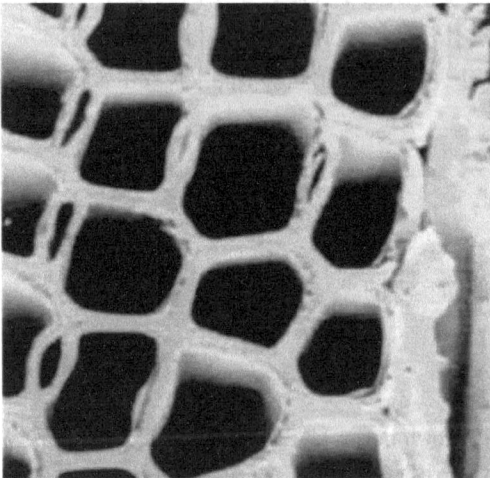
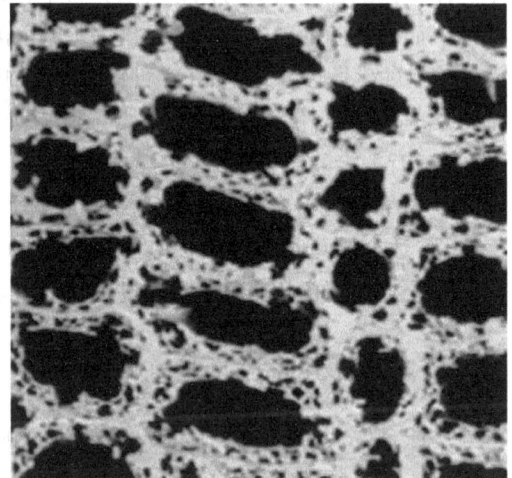

图 8.10　受软腐菌侵蚀的雪松

8.2.3　木材变色

　　树木伐倒或制成板材后，木材正常颜色发生改变的，叫木材变色（discoloration）。按照木材变色的起因，木材变色可以分为：化学变色、腐朽菌变色、霉变色和变色菌变色（图 8.11）。木材化学变色是木材锯解后由于化学或生物化学反应引起的浅棕红色、褐色或橙黄色等不正常的颜色，其颜色一般比较均匀，且分布仅限于表层。后 3 种变色属于真菌引起的木材变色。

　　◎ 化学变色：锯材由于化学或生物化学反应引起木材变色；

　　◎ 腐朽菌变色：木腐菌引起的木材变色；

　　◎ 霉菌变色：霉菌引起的木材变色；

　　◎ 变色菌变色：变色菌引起的木材变色。

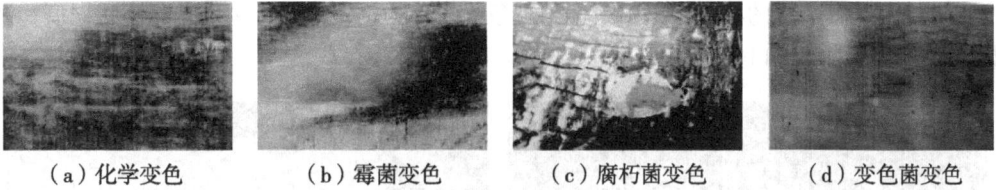

（a）化学变色　　　　（b）霉菌变色　　　　（c）腐朽菌变色　　　　（d）变色菌变色

图 8.11　木材变色

8.2.4　木材虫害

因各种昆虫危害而造成的木材缺陷称为木材虫害，常见的害虫有小蠹虫、天牛、吉丁虫、象鼻虫、白蚁和树蜂等，其中白蚁对树木或木制品的危害是非常严重（图 8.12）。木材虫害主要有表面虫眼和虫沟、小虫眼、大虫眼之分（图 8.13）。

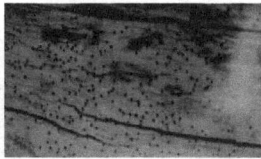

表面虫沟　　　　　　　　小虫眼　　　　　　　　虫及大虫沟

图 8.12　白蚁　　　　　　　　　图 8.13　木材虫害

8.3　木材加工缺陷

木材在加工过程中所造成的木材表面损伤，称为木材加工缺陷。有锯解缺陷与干燥缺陷之分。在木材锯解过程中，主要会产生缺棱（wane）或锯口缺陷两种锯解缺陷。木材在干燥中或干燥后所产生的缺陷称为干燥缺陷。

缺棱（wane）指在整边锯材中残留的原木表面部分（图 8.14），分为钝棱（wanting arris，wane）和锐棱（wanting edge）。锯材宽度、厚度方向的材棱未着锯的部分，称为钝棱。锯材材边长度未着锯的部分称为锐棱。

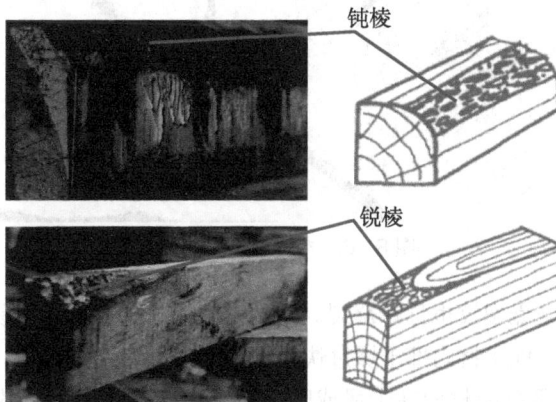

钝棱

锐棱

图 8.14　锯材缺棱

木材因锯解造成的材面不平整或偏斜现象，统称为锯口缺陷，主要有瓦棱状锯痕、波状纹、毛刺糙面和锯口偏斜四种（图 8.15）。

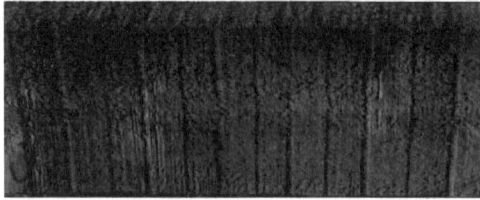

瓦棱状锯痕

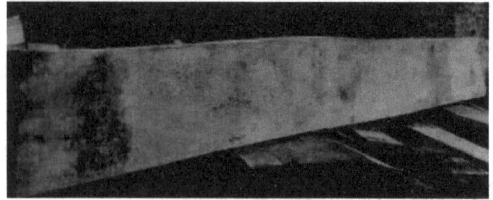

波状纹

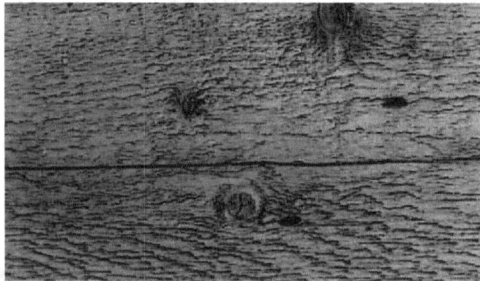

毛刺糙面

锯口偏斜

图 8.15　锯口缺陷

木材干燥过程中或干燥后容易产生干燥缺陷，主要包括开裂（split）和翘曲（warping）。

开裂分为：端裂（end check）和内裂（honeycomb check）（图 8.16）。端裂出现在木材的端部，是沿木射线方向产生的径向开裂。产生的原因在于木材顺纹导水性远远大于横纹方向，干燥时水分从端面蒸发比从侧面蒸发快得多，端部含水率低于中部，端部的收缩受中部木材的限制，从而在端部产生了拉伸应力，当拉应力大于木材的横纹抗拉强度时，就产生了端裂。

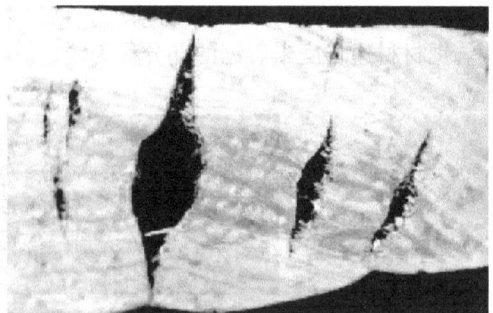

图 8.16　木材端裂和内裂

木材内部密集的干燥裂纹也称蜂窝裂。内裂不到达木材表面，不易发现。这是由于木材内层拉应力引起的。即木材内部干燥过快，大于材表干燥速度所致。

木材翘曲指的是平整的材面在干燥或贮存后，出现材面不平整的现象。这是因为木材的各向异性、干燥不均匀或者堆垛不良造成的。翘曲分为：扭曲（twist）、翘弯（cup）、顺弯（bow）和横弯（crook）（图 8.17）。

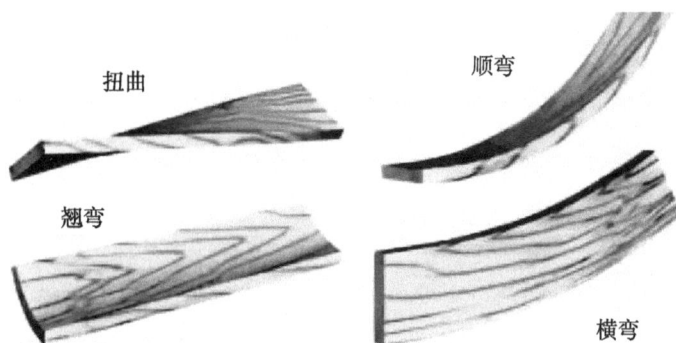

图 8.17　木材翘曲

▶▶ 思考与训练

1. 什么是节子？如何形成？如何区分活节和死节？

2. 什么是应力木？对加工有何影响？

3. 真菌的生长条件有哪些？

4. 木材腐朽的类型及其微观和宏观表现？

5. 木材加工缺陷有哪些？

Chapter ❶ »»»
Forest and Wood

As the most complex, functional, and stable terrestrial ecosystem on Earth, forests act as the general regulators of nature and the lungs of the Earth. Forests play a decisive and irreplaceable role in maintaining ecological balance, promoting harmony between man and nature, and protecting human survival and development.

In addition to playing an important ecological function, forests directly provide abundant *wood* materials for human living. Among the four key building materials, i. e. *cement*, *steel*, *wood*, and *plastic*, wood is the only renewable ecological material. Corresponding with the advancements of modern industry and technology, wood use has been widely adopted.

However, wood is often not cherished because of its prevalence and easy obtainability. Long-term brutal and unreasonable exploitation has resulted in a sharp decline in forest resources, as well as unprecedented ecological disasters. Although people have realized the close relationship between human survival and forest ecosystems, there is still a lack of understanding towards the preciousness of wood and its rational usage. To protect forest resources and employ rational uses of wood, it is necessary to maintain a proper reverence and value towards the natural world.

1.1 Forest and Human Survival

1.1.1 State of Forest Resources

There are many different kinds of forests. Generally, forests can be divided into two types: natural forests and planted forests. According to their functions, social demand, and management purposes, existing forests can be divided into five types, namely, timber forests, shelter forests, economic forests, firewood forests, and special-use forests (Fig.1.1).

The defining criteria for forests are as follows:

◎ The *canopy density* is above 0.2;

◎ The natural forest area occupies more than 1.5 mu (about 1000m^2), while planted forests and economic forests occupy more than 1 mu (about 667m^2).

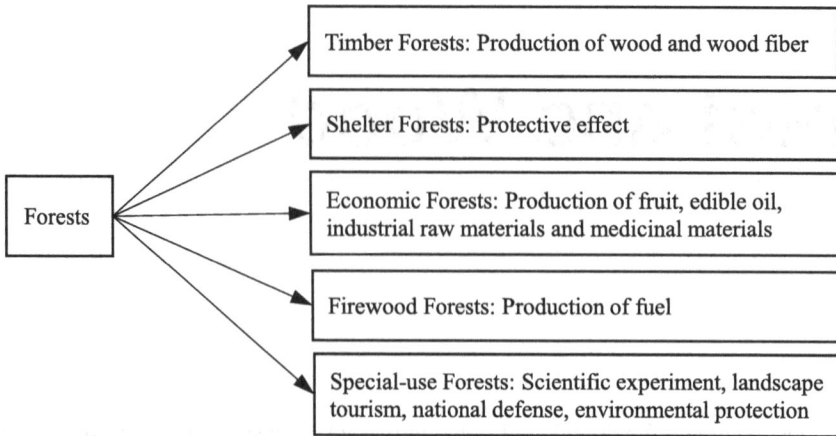

Fig.1.1　Forest classification for the given use

Globally, forests occupy approximately 4.06 billion hm^2, of total area, with a coverage rate of around 30.8%, for which coniferous forest accounts for one-third. However, the distribution of global forests is extremely unbalanced. The top five countries with the most abundant forest resources (Russia, Brazil, Canada, the United States and China) account for more than half of the total forest area in the world. Natural forests represent 93% of the world's forests, with the largest section (45%) located in the tropical zone, followed by the sub-frigid zone, temperate zone, and subtropical zone.

According to the 9th China Forest Resources Inventory (2014—2018), the existing forest area in China is 220 million hm^2, with a forest volume of 17.56 billion m^3, and a forest coverage rate of 22.96%. China, as a country, hosts the most forest tree species in the world, especially in terms of rare and precious species. There are more than 20,000 species of seed plants in China, including more than 8,000 species belonging to forest trees. Among these species, more than 2,000 kinds are used for arbors alone. There are also more than thousands of species demonstrating favourable properties, such as tall, straight trunks and high economic value.

Pine and *fir*, as coniferous tree species, are the main species dominating the northern hemisphere. There are about 30 genera in the world. In China exclusively, there are 20 genera and nearly 200 species. Among them, eight genera are unique to China, namely, *Metasequoia*, *Cathaya*, *Larix*, *Metasequoia*, *Taiwania*, *Keteleeria*, *Fokienia* and *Cunninghamia*.

Globally, the types of broad-leaved trees are even more diverse, with over 200 genera. Many of these are endemic to China, such as *Davidia involucrata*, *eucommia*, *Camptotheca*, *fragrant fruit tree* and *cherry pepper*.

1.1.2　Shortage of Forest Resources in China

In comparison with the other countries and regions in the world, the main deficiencies of China's forest resources are as follows:

(1) Few Forest Resources and Low Coverage

China's forest coverage rate is only about two-thirds of the global average level. The forest area per capita is less than one-fourth of the global average level, whereas the forest volume per capita is only one-seventh of the global average.

(2) Unbalanced Distribution of Forest Resources

The vast majority of forest resources in China are concentrated in remote mountain areas. In contrast, the vast northwest in China lacks forest resources.

◎ The northeast forest area is the largest natural forest area in China, with forest resources mainly concentrated in Daxing'an Ling, Xiaoxing'an Ling and Changbai Mountains. The main tree species include *Korean pine, Larix gmelinii, Larix gmelinii* and other coniferous trees, as well as *birch* and *Northeast China ash* belonging to the broad-leaved tree species.

◎ The southwest forest area is the second largest natural forest area in China, mainly located in Hengduan Mountains. The main tree species include *spruce, fir, Quercus Alpinia, Pinus yunnanensis,* and *precious teak, red sandalwood, camphor,* etc.

◎ The southeast forest area is the main economic and artificial forest area in China. It features warm climates, abundant rainfall, and favourable plant growth conditions. There are many kinds of trees, mainly *Cunninghamia lanceolata* and *Pinus massoniana*, and unique bamboo forests in China.

(3) More Timber Forests than shelterbelts

The proportion of forest species is unevenly distributed across China. Therefore, it is difficult to obtain the multiple, comprehensive benefits of forest resources.

(4) Low Quality of Forest Resources

The proportion of forest land within China's forestry is low, including forest volume per unit area and forest growth rates. Among the existing forests in China, there are few original forests, instead there lies more natural defective forests and different forest types. There are few recoverable forest resources, and the contradiction between supply and demand for wood is prominent.

1.1.3 Ecological Functions of Forests

(1) Oxygen Production and Carbon Fixation

Every tree is an *oxygen generator* and *carbon dioxide absorber*. Forest plants absorb carbon dioxide and release oxygen through photosynthesis, and fix carbon dioxide from the atmosphere in the form of vegetation and soil biomass. Findings demonstrate forests absorb 1.83 tons of carbon dioxide and releas 1.62 tons of oxygen per 1 m^3 of forest area.

Forests are the largest carbon storage and the most economical carbon sink on land (Fig.1.2). More than half of the carbon stored in the terrestrial ecosystem derives from the forest ecosystem, with the global forest carbon storage reaching 289 billion tons. Despite forest areas accounting for one-third of the global land area, their annual carbon dioxide absorption accounts for four-fifths of the total biological carbon fixation.

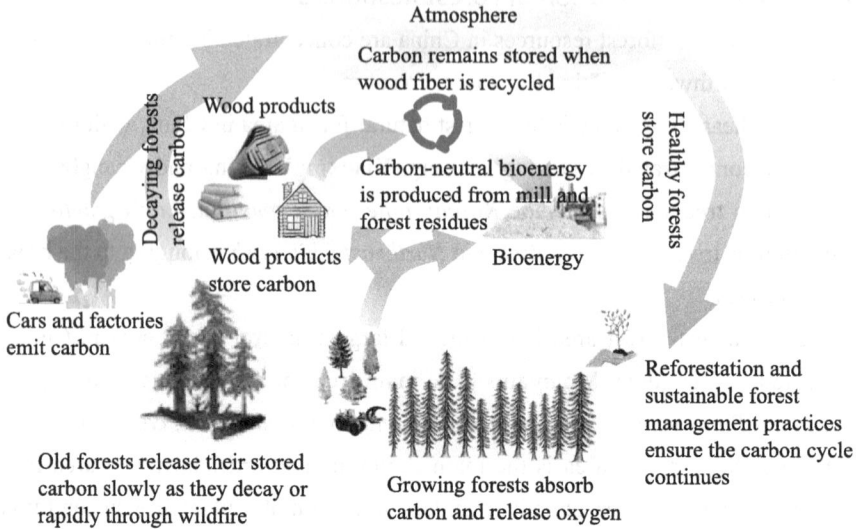

Fig.1.2　Sustainable forestry carbon cycle

The amount of carbon fixed in the forest can be stored for a long time, and can also be found in various forest products. In turn, forest carbon fixation is one of the most economic and effective ways of dealing with climate change. Therefore, a global consensus and action has developed within forestry, i.e. to accelerate the cultivation of forest resources and enhance the functions of forest carbon sinks.

(2) Temperature and Humidity Regulation

Forests are natural *air conditioners*. Under the dense crown canopy, green thermostats are formed between the Earth's surface and atmosphere. These green thermostats not only produce special changes in the forest, but also have a significant impact on the temperatures surrounding the forest. Compared with non-forest lands, forests are typically warmer in the winters and cooler in the summers. Forest nights are warmer, whereas days are cooler, with only slight differences in temperature.

(3) Soil and Water Conservation

Forests have complex multi-layer structures and adequate surface coverages. These include strong plant roots and soil infiltration systems that effectively intercept surface runoff from precipitation to prevent and reduce soil and water loss.

Forests redistribute and purify precipitation through canopy interception, forest litter water holding, and soil regulation. Not only do these processes conserve and store water, but also reduce the hardness of water to improve water alkalinity and quality (Fig.1.3).

Fig.1.3　Water conservation

Fig.1.4　Soil improvement

(4) Soil Improvement

Tree roots in the forest are rich, and the penetration of these roots improves the physical structure of the soil. Under the actions of wind, precipitation, sunlight, microorganism, and various animals, forest land litter, dead ground cover, and animal carcasses can be decomposed into *humus*. These decomposed remains improve the contents of soil organic matter such as nitrogen, phosphorus, potassium, and other elements required by plant growth altogether improving and increasing soil fertility. At the same time, feces produced by birds and animals also play a great role in maintaining the fertility of forest soils. Plants absorb cadmium, lead, copper, zinc, mercury, and other heavy metals found in the soil through their roots, reducing the pollution caused by heavy metal in soil (Fig.1.4).

(5) Sterilization and Disinfection

Forests can secrete *fungicides* such as terpene, alcohol, organic acid, ether, aldehyde, and ketone. These substances can kill bacteria, fungi, and protozoa, while significantly reducing the number of bacteria among forest air. According to monitoring efforts, one hm^2 of broad-leaved forest can produce up to 2kg of phytobactericide within a 24-hour cycle, whereas one hm^2 of coniferous forest produces than 5kg.

(6) Purified Air

Many plants can absorb pollutants in the atmosphere under the condition of maintaining normal physiological functions. The metabolizing, degrading or enriching of pollutants in the body allows, the air to be purified to a certain extent. Sulfur dioxide is one of the main air pollutants. Sulfur is a amino acid component found in trees, and it is also one of the nutritional elements needed by trees.

(7) Dust Retention

There are a lot of dusts floating in the atmosphere, which is the main sources of air pollution. These usually include industrial dust, carbon particle, gasification iron, manganese, zinc, asbestos, cement, and fiber powder. Dust retention effects of the forest are shown as follows:

◎ On the one hand, dense branches and leaves of trees can block out air flow and reduce wind speed. With decreases in wind speed, smoke and dust will lose their momentum to move through the atmosphere, instead falling to the ground.

◎ On the other hand, tree leaves have a strong transpiration surface. On sunny days, they need to transpire a lot of water to keep the humidity of the canopy and forest surface relatively high. Subsequently, smoke and dust absorption can accordingly increase. As the adsorption capacity of tree leaves increases, smoke and dust particles are easier to fall and be adsorbed. However, on rainy days, the smoke and dust of tree leaves are washed away by the rain. Thus, after repeated rainfall in the forest, the air will become clean.

◎ Finally, the flowers, fruits, leaves, and branches of trees secretes a variety of sticky juices through their rough and hairy surfaces. In turn, dust and smoke in the air adheres to the concave parts of leaves and branches after passing through the forest, thus playing role in adhesion, blocking and filtering.

(8) Noise Isolation

The rough trunk, dense branches and leaves of forest trees alongside the litter layer can reflect, absorb and block noise (Fig.1.5). A 40-meter-wide arbor belt can reduce noise by 10~15 dB, up to 30dB, whereas a 20-meter-wide arbor belt can reduce noise by 8~10 dB. Greenways in the city can reduce noise by 3~10 dB more than those without greening. Meanwhile, trees in the park can reduce noise by 26~34 dB.

Fig.1.5　Noise isolation　　　　　Fig.1.6　Forest medical care

(9) Medical Care

In forests, photoelectric effects can be formed through the electric discharge from trees and branches, as well as photosynthesis from green plants. These processes ionize the air and produce negative ions. The negative air ions, also known as "air vitamin", have a strong inhibitory effect on various bacteria and viruses. These negative ions can neutralize harmful substances in the air and inhibit the growth of bacteria and also enter the human blood through the respiratory tract by passing through the lungs. Thus, they promote blood circulation, stimulate the central nervous system, regulate people's emotions, and improve human immunity. Therefore, air ions have excellent physiological effects on the human body and provide medical benefits (Fig.1.6).

(10) Maintaining Biodiversity

Forests are widely distributed in a wide range of natural and geographical environment types, which are the most complex and stable terrestrial ecosystems on Earth. On the one hand, forests provide the base survival and nutritional source for various plants and microorganisms. On the other hand, they provide habitat and rich food for animals, that are suitable for the survival and reproduction of many species (Fig.1.7).

Scientists have estimated that human beings need about 40,000 living species to maintain their everyday clothing, food, housing, and transportation. Scientists have observed that the disappearance of a single species can cause the disappearance of 20 associated species. Therefore, we should protect nature, protect forests, and maintain biodiversity.

Fig.1.7　Forest biodiversity maintenance

1.1.4　New Direction of Forest Development and Utilization—Forest Health

In forest environments, people participate in climbing, hiking, sightseeing, hunting, exploring, camping, recuperation, and other activities (Fig.1.8 and Fig.1.9). According to statistical data, National Forest Parks are the most prosperous places of forest tourism among developed countries. The world tourism organization has redicted that in the 21st century, forest tourism as the main form of eco-tourism, will be the fastest growing part of the tourism industry, with a growth rate of 30%. More than half of global tourists will choose to enter and enjoy forest environments.

Fig.1.8　Forest hiking

Fig.1.9　Forest recuperation

1.2　Wood and Human Living

In human history, wood has accompanied the evolution of human beings. Our ancestors created fire with wood, made artifacts with wood, and built houses with wood, and gradually established human civilization. It can be said that wood is essential in people's clothing, food, housing, and transportation. Even today, wood still plays an irreplaceable role in daily life.

1.2.1　Wood Culture

The history of human development also corresponds with the history of wood utilization. In Chinese history, most of the buildings, bridges, furniture, boats, carts, weapon parts, musical instruments, coffin sets, religious instruments, study utensils, toys, chess and cards, fishing tools, bird and insect cages are all made of wood. These implements reflect and inherit the extensive and profound Chinese culture.

Wood culture is formed in people's life, which are related to wood culture, shared wood values, and wood utilization methods (Fig.1.10). Wood cultural and creative products are the best interpretation of contemporary "wood culture", including two categories of wood crafts (Fig.1.11) and derivative cultural and creative products (Fig.1.12-Fig.1.14). Wood handicrafts represent the historical, cultural, and regional features through the use of traditional artisanship such as sculpture, inlay and hot stamping, or modern techniques to improve the design, that better integrates the culture of handicrafts into contemporary society. Derivative cultural and creative merchandise is practical, aesthetic and interesting. They mainly include cultural lifestyle products, cultural tourism products, and media products. The cultural undertone of wood materials can elevate the connotation of products. Furthermore, the sensory experience of wood materials can generate emotional resonance.

Fig.1.10　The evolution of Chinese character of "wood"

1.2.2　Wood Architecture

Wood has been widely used in Chinese traditional architecture. Wood architecture is the most outstanding carrier of Chinese civilization in the aspect of living. From the ancient ancestors' burrowing, rammed earth houses, to the beautiful and improving cornices and corners, carved beams and painted buildings, ancient Chinese architecture has been unique for thousands of years thanks to the Chinese wisdom and aesthetic sustenance. There are many kinds of ancient buildings, including palaces, temples, pagodas, garden buildings and residential buildings. Nanchen temple is the oldest existing wood construction of the Buddha Hall (Fig.1.15). Beijing forbidden city (Fig.1.16) is one of the largest and most well preserved ancient wooden buildings in the world. Shiga pagoda (Yingxian Wood Pagoda) of Fogong Temple in Yingxian County, Shanxi Province, built in Liao Dynasty (Fig.1.17), is the only existing wood pagoda in China and a prime example of ancient wood high-rise buildings.

Fig.1.11   Phoenix fragrant holdings

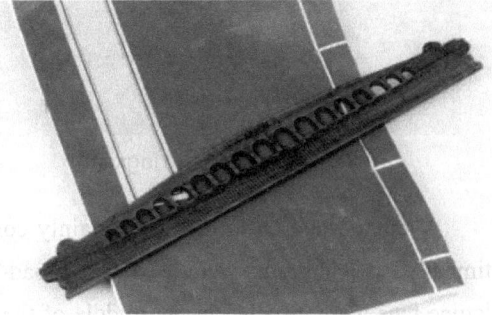
Fig.1.12   17-hole-bridge ruler

Fig.1.13   Wood bookmark

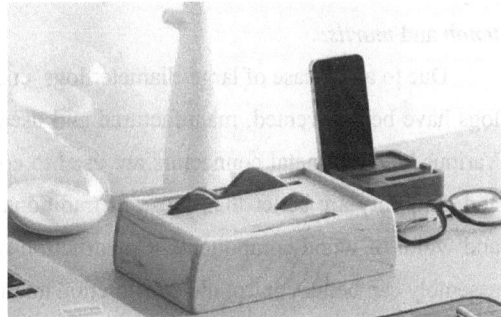
Fig1.14   Tissue box

Fig.1.15   Grand buddha hall of Nanchan temple

Fig.1.16   Taihe hall of the Forbidden city

Fig.1.17   Pagoda in Yingxian

Fig.1.18   Bracket set

Ancient wood buildings were mainly constructed using the post-and-beam systems. The timber posts and beams were used as the load-bearing skeleton. The walls, doors, windows, and lattice fans were installed in the middle of the column according to their use requirements. The beam frame was used to support the big roof, and the paving layer, namely *bracket set* (" 斗拱 "), played a transitional role between the posts and beams (Fig.1.18). The joints were combined with *tenon* and *mortise*.

Due to a decrease of large-diameter logs, engineered wood products made from small-diameter logs have been invented, manufactured and used to construct modern timber buildings (Fig.1.19). Various kinds of metal connectors are used to connect wood components. Their structural stability and durability are better than those of traditional wood fasteners, which expands the span, height and width of wood components. Construction methods with a high degree of prefabrication and assembly can better control the construction quality while speeding up construction.

1.2.3   Wood Furniture

Chinese traditional furniture alludes a gorgeous poem created by craftsmen with wood. Especially, furniture from the Ming and Qing dynasties are treasures of Chinese traditional culture, representing the most traditional Chinese style of classical furniture.

Fig.1.20 (a) shows one classic Ming-style armchair. The structure of the armchair is round outside and square inside, with a round top and square bottom. The back board is designed into a "S" or "C" shape curve according to the curve of the human spine. It is a typical example of the scientific nature of Ming-style furniture.

The new Chinese-style furniture [Fig.1.20 (b)] uses modern materials and craftsmanship to portray the essence of traditional Chinese culture. This furniture not only has elegant and dignified Chinese element, but also demonstrates apparent modern characteristic.

（a）Light wood-frame construction

（b）Log construction

（c）Post-and-beam wood construction

（d）High-rise wood construction

Fig.1.19　Modern wood buildings

Fig.1.20 (c) shows the Louis XV armchair, which is a typical Rococo-style. It is composed of smooth curves with its armrests oriented outwards and retracted inwards. The back and seat are covered with elegant and beautiful brocade. The foot ends are often carved into cat's or goat's feet, which are delicate and beautiful.

The three-legged shell chair is shown in Fig.1.20 (d). Because its chair surface is graceful and looks like a warm smile, the chair is also called "smiling chair". Its seat surface presents a unique three-dimensional curved surface effect, which is as light and smooth as a feather, floating nimbly.

1.2.4　Wood Decoration

In the decoration of buildings, wood, a natural ecological material, has been widely used. These include floorings and ceilings in a horizontal direction, with wall surface, column surface, doors, windows and wood lines in a vertical direction (Fig.1.21-Fig.1.23).

（a）Classic Ming style armchair

（b）The new Chinese style furniture （c）Western classical furniture （d）Western modern furniture

Fig.1.20 Typical furniture in ancient and modern China and abroad

（a）Kindergarten （b）Basketball hall

Fig.1.21 Wood flooring for ground decoration

Fig.1.22 Wood for interior ceilings and walls

(a) Harbin grand theatre, China

(b) Wild reindeer lookout pavilion, Norway

Fig.1.23 Wood for interior decoration of buildings

1.2.5 Wood for Medicine

Many kinds of wood have high medical value. The medical use of wood has been present in traditional Chinese medicine for thousands of years, and is still in use today (Fig.1.24). Scented rosewood (*Dalbergia odorifera*) from Hainan is an extremely precious wood species. The "*odorifera*" extracted from the heartwood has anti-tumor, cardiovascular protection, anti-inflammatory, anti-oxidation of free radicals and sedative effects. The oil extracted from *Dalbergia odorifera* after distillation can be used as analgesics, and it is used in combination with other medicines in traditional Chinese medicine. It is known as an antihypertensive wood because of its antihypertensive effect.

Fig.1.24 Traditional Chinese medicine pharmacy

1.2.6 Wood Energy

With the depletion of traditional *fossil energy* and the increasingly severe energy crisis, the development of renewable energy is imminent. Since the time of drilling wood for fire, wood is the earliest energy resource, being both renewable and degradable. It is expected to become a new way to solve the modern energy crisis. Use of wood for energy is the most effective means. With modern science and technology, wood can be processed into an efficient and convenient form of energy. The following three methods of wood energy processing are commonly used: (1) *wood liquefaction* to produce fuel oil; (2) *wood hydrolysis* to produce fuel ethanol; and (3) *wood pyrolysis* and *gasification* for power generation.

Wood is a large molecule polymer composed of *carbon, hydrogen* and *oxygen*. These chemical elements are important components of fuel materials. Wood can be liquefied into fuel oil by physical, chemical, or biological methods. Common methods of wood liquefaction, to convert wood into biodiesel, include the high-temperature and high-pressure liquefaction, atmospheric pressure catalyst liquefaction, supercritical liquefaction, and microbial assisted liquefaction.

Wood is mainly composed of three major polymers, namely *cellulose, hemicellulose* and *lignin*. Among them, cellulose and hemicellulose are mostly sugars, accounting for 75% of the total wood composition. The key to ethanol production from wood is the hydrolysis of high glycan in wood, and the main ways are through enzymatic hydrolysis and acid hydrolysis (Fig.1.25).

Under a high temperature, wood materials can be converted into a combustible gas under the action of a gasification agent (Fig.1.26). Then, the purified combustible gas can be sent to an internal combustion engine to directly generate electricity. The first sawdust gasification power plant in China was built in Sanya, Hainan in 2000. The power plant uses wood wastes as raw materials and has achieved good energy-saving and environmental benefits.

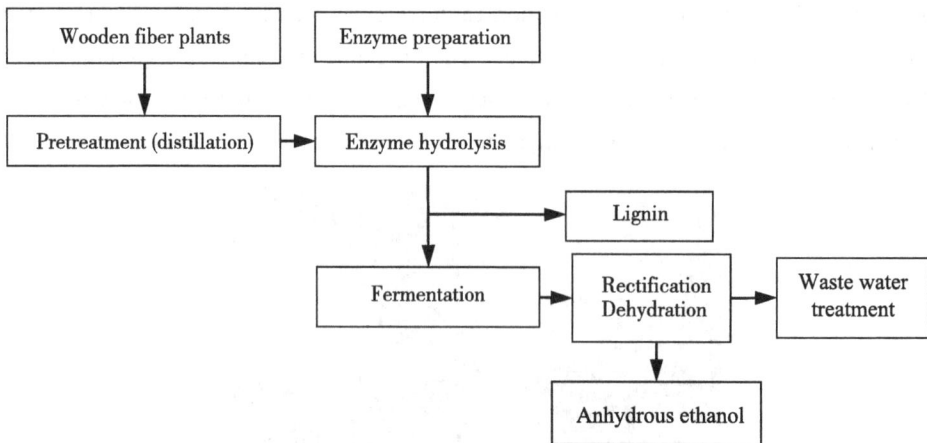

Fig.1.25　Process of wood tofuel ethanol

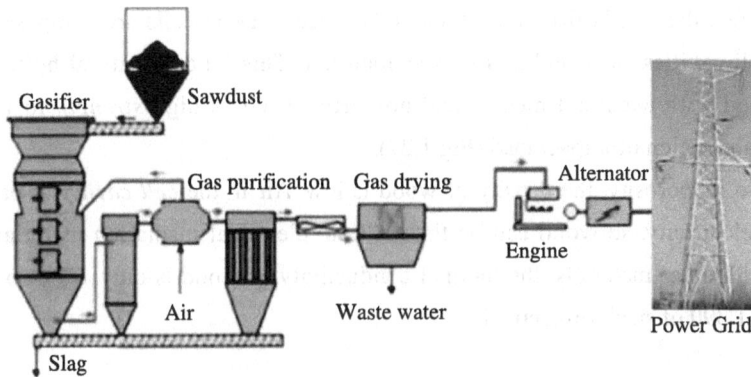

Fig.1.26　Process of sawdust gasification power generation

The goal of protecting wood resources and the ecological environment has become an important topic to ensure energy security. In China, there are 154 kinds of woody oil tree species with an oil content of more than 40%. In turn, there are about 300 million tons of energy that can be used to generate electricity by burning branch residues every year. Diesel oil is extracted from the fruits or seeds of trees, and the energy engine is solved by burning wood fiber for power generation. Presently, biomass energy developed by Jatropha curcas and Hippophae rhamnoides has become the hot topic and trend of new energy development.

1.2.7　Pulping and Papermaking

Paper is an indispensable object in people's cultural and daily lives. Wood *pulping* and *papermaking* is one of the essential ways towards a comprehensive utilization of forest resources, as well as an important processing method to achieve a high value added to wood. Pulping is a process that uses mechanical or chemical methods, or a combination of the two, to separate the fibers of wood raw materials.

1.3　Characteristics of Wood

1.3.1　Biological Characteristics of Wood

The figure of wood with its annual rings as the main body has different patterns in different sections, giving people a feeling of fluency, relaxation, coordination, and elegance. The rings of different width represent the feeling of dialogue with nature. It fluctuates with the growth environment, consistent with the inherent fluctuation of biology. It is easy to arouse people's psychological resonance, and can also give people a sense of security. Therefore, wood is the closest and most humane material to human beings.

1.3.2　Porosity of Wood

Wood consists of cells produced during the growth of trees. These cells in wood have become lignified dead cells from the living cells in trees. Dead cells are composed of the cell walls and cell cavities, so wood is a *porous* material. This kind of natural hollow structures provides wood with excellent mechanical properties such as high *strength to weight ratio*, rigidity, and absorption of impact load (Fig.1.27).

Owing to its porosity, the density of wood is low. Air in the *cell cavityis* a bad conductor of heat and electricity, so wood can be thermal and electrical insulation material (Fig.1.28). Compared with other materials, the thermal conductivity of wood is only 1/6 of brick, 1/15 of concrete and 1/390 of steel, respectively.

Fig.1.27 Wood container for packaging precision instrument   Fig.1.28 Cork used for thermos bottles

The porous tubular structure of wood renders excellent sound amplification and resonance performance (Fig.1.29). *Paulownia* and other wood species are selected as sound materials for many musical instruments.

Fig.1.29 Musical instruments

The *porosity* of wood makes the wood have special surface properties. For instance, surface reflection and inner reflection of light give the wood a certain luster and softness. Furthermore, wood enables ultraviolet light absorption that can be harmful to eyes and resultingly cause glare. Hence, the special surface properties of wood should be considered while painting and gluing.

Porosity increases the specific area of wood. Activated wood charcoal makes use of the porous characteristics of wood (Fig.1.30). Different from other materials, porosity

Fig.1.30 Activated wood charcoal

makes wood have gas and liquid *permeability*. Thus, permeability should be fully considered when wood is dried, preserved, and modified.

1.3.3 Anisotropy and Variability

The growth of trees is formed by the *cambium* between *bark* and *xylem*. The cambium thickens year by year, forming a whorl-like cylinder. There are transverse ray cells from the pith center to the bark which form radial *wood ray tissue*. Therefore, wood has nearly cylindrical symmetry with three basic directions: *longitudinal, radial* and *tangential*. The moisture transfer, shrinkage, heat, electricity, sound, and strength of wood are different in these three different directions.

As a biomaterial, wood has a noticeable variability besides anisotropy. Wood properties vary with tree species, origin, and location. The variability of wood makes it difficult to be called a real engineering material. The variability of wood-based materials can be effectively reduced by element recombination.

1.3.4 Durability

Wood is a high-molecule compound composed of *cellulose, hemicellulose, lignin*, and *extractives*. These components contain *hydrophilic hydroxyl groups* with hygroscopicity, which affects dimensional stability. They are easy to decompose under high temperature and become flammable materials. Under a certain humidity, they can provide nutrients for bacteria and insects, causing wood discoloration, decay, and insect damage. At the room temperature, the main components of wood are stable in nature and do not react with most chemicals. Therefore, wood can be used for applications with high chemical resistance requirements.

1.3.5 Renewability

Wood is a kind of biomaterial, belonging both to renewable and new energy materials at the same time. Based on the existing worldwide forest volume, forests can increase by at least 3 billion m^3 per year. These demonstrates obvious advantages over underground resources, with oil predicted to be exhausted in the 21st century. Taking into account of the current usage of mineral resources, *silver, mercury* and *zinc* can only be exploited for another 20 years; *tin* and *lead* for 40~50 years, *copper* and *nickel* for 60~80 years and *iron* and *manganese* for 160~170 years. As long as trees are planted, wood can become an inexhaustible renewable resource and energy.

Exercises

1. Make a brief introduction about global forest resources.
2. Summarize the forest resources available in China and indicate its key characteristics.
3. Explain the eco-functions of the forest.
4. Justify why wood is called a sustainable material.

Chapter ❷ »»»
Sources of Wood

Trees are complex organisms. Originating either from *vegetative propagation* or from *sexually fertilized eggs* that become tiny *seed-encased embryos*, trees grow to be one of nature's largest living organisms.

Wood itself stems from plants. However, not all forest plants can produce wood. Wood can be understood both in narrow sense and broad contexts. In a narrow sense, wood refers to the *xylem* of a 4~6m tree trunk. In a broader sense, wood refers to wood materials, including not only the poles and logs produced from forest harvesting, but also wood-based products, such as *sawn timber, plywood, particleboard*, and *fiberboard*. This chapter will introduce the formation and nomenclature of wood, alongside with information about tree growth.

2.1 Plant Taxonomy and Wood Products

Plant taxonomy is a basic discipline that mainly studies the origins, relationships, and evolution of different groups throughout the plant kingdom. In other words, the complex plant kingdom has been classified into species and arranged in a systematic way to facilitate the understanding and utilization of plants.

Engeler's natural classification is often used in plant taxonomy and is based on the main morphological characteristics of plants (which are flowers, fruits and leaves). The commonly used taxonomic units for plants are kingdom, phylum, class, order, family, genus and species. Species are the basic taxonomic unit and are delineated by hybrid sterility. Closely related species are grouped into genera; closely related genera into families; closely related families into orders; closely related orders into classes, and so on. For example, the taxonomic classification of *Cypressus funebris* is as follows.

Plantae is divided into four categories: *algae, bryophytes, ferns*, and *seed plants* (Fig.2.1). Among these, seed plants hold the most species and the most complex structure of all plants on Earth. There are 200,000~250,000 species in the world and 30,000 in China. Seed plants include woody and herbaceous plants. In turn, this plant category is closely related and actively involved in human economic life.

Kingdom	**Plantae**
Phylum	Supermatophyta
Subphylu	Gymnospermae
Class	Coniferae
Order	Coniferales
Family	Cupressaceae
Genus	Cupressus
Species	Cupressus Funebris

(a) Algae

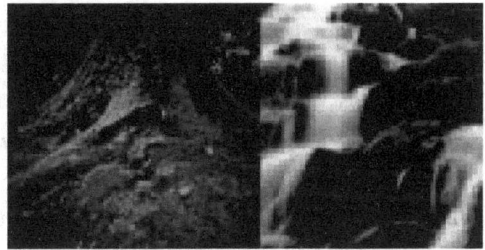

(b) Bryophytes

(c) Ferns

(d) Seed plants

Fig.2.1 Four categories of plants

Trees belong to the seed plants category, which is a general name for woody plants, including tall perennial trees, low tufted shrubs and vines that twist around other structures. Wood mainly comes from the trees of gymnosperms in seed plants and dicotyledons in angiosperms.

If classified from the perspective of wood products, wood can be divided into standing trees, felling woods, raw poles, logs, sawn lumber, and wood-based panels (Fig.2.2).

◎ *Standing tree*: trees grown in woodlands.

◎ *Felling wood*: trees that are felled.

◎ *Poles*: felling trees by removing branches.

◎ *Logs*: segments of a certain size cut from poles.

◎ *Lumber*: parts sawn from logs in the length direction. Sawn wood can be divided into timber and lumber. Sawn wood with a width/thickness of less than 2 is called timber, and sawn wood with a width/thickness of larger than 2 is called lumber.

◎ *Wood-based panels*: panels made of wood fibers, strands or veneer and an adhesive(s).

Standing Tree   Felling Tree   Pole   Log   Timber   Lumber

Wood -based panel

Fig.2.2　Wood products

2.2　Tree Growth and Wood Formation

Both *broadleaves* and *evergreens* start life as *seeds* on the ground, and they go through a life cycle as plants. Some trees grow and mature more slowly, and some trees will live for centuries before they eventually die. Many trees grow to incredible sizes. For example, some of the redwood trees in the United States of America have archways through their trunks that are large enough to drive a car through (Fig.2.3).

Fig.2.3　A car driving through a redwood tree

The mechanisms that result in tree growth are remarkably complex, which is becoming more and more evident with advances in science. In the following paragraphs, trees' growth is discussed in rather general terms to provide a basic understanding of the processes involved. Wood (*xylem*) is found inside a covering of bark, which is composed of an inner layer (*phloem*) and an outer protective layer (*outer bark*). As a tree grows, it adds new wood, increasing the diameter of its main trunk and branches. Bark is also formed in the process of growth while it cracks and flakes off the trunk at the same time.

Similar to other plants, tree seeds are planted by many approaches. Some seeds simply drop from a tree to the ground, while wind, animals, and birds may carry others. Seeds can also be planted in pots or trays in nurseries. After some time, the seed germinates, and the shoot (*plumule*) that will form the trunk makes its way towards the heat of the Earth's surface (Fig.2.4). The sapling continues to grow and eventually becomes an arbor with luxuriant foliage and well-developed roots.

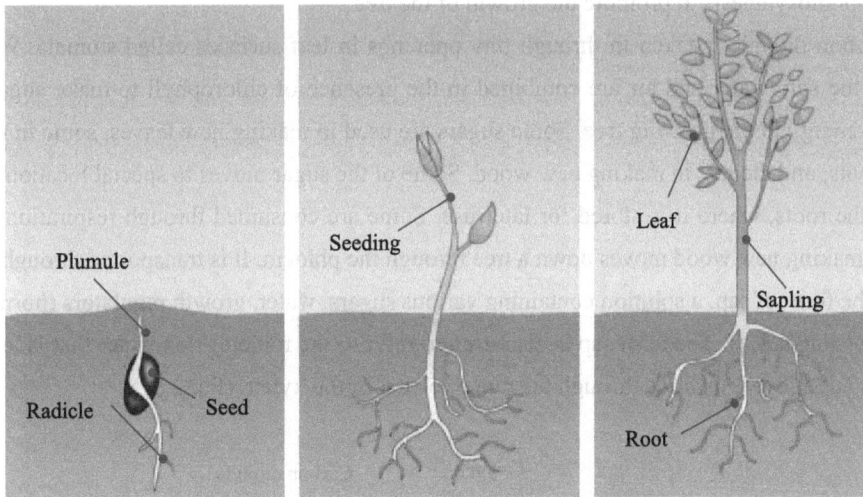

Fig.2.4 The stages in the growth of a tree

As young trees grow, the leaves begin to use energy from the sun to convert *sap* into food. Trees do this using a substance in their leaves called chlorophyll. This process is called photosynthesis. It needs only water (from the soil), carbon dioxide (from the atmosphere), and light (from the sun) to do this. Water, along with nutrients, move from the roots, through the outer part of the xylem, and is sent up to the leaves. Wood cells provide pathways for the unbroken fluid columns that link the roots to the leaves. It is worth noting, the capillary action is, based on the laws of physics, influenced by pore diameter and surface tension. Taking into account of these factors, a tree can grow as high as 116m.

Roots are the underground component of a tree, comprising of a main root, side roots and fibers. They account for 5%~25% of the volume of a *standing tree*. The main root supports the tree, anchoring the latter to the ground and ensuring its growth. The side roots and fibers mainly take up water and nutrients from the soil to the leaves for *photosynthesis*.

The *trunk*, which is the main body of the tree, is the upright section between the crown and the root. It accounts for 50%~90% of the volume of a standing tree. In a standing tree, the trunk has three important functions: transportation, storage and support. The *sapwood* of the xylem transports water and minerals absorbed by the roots to the crown. Then, the organic nutrients produced by the *crown* are transported throughout the tree, via the phloem of the bark, and stored in the trunk.

The *crown* is a general term for the branches, leaves, lateral buds and terminal buds on the top part of a tree, accounting for 5%~25% of the volume of the standing tree. The scope of the crown is usually calculated from the first large live branch at the top of the trunk to the top of the crown. The branches in the crown transport the nutrients absorbed by the roots from the sapwood to the leaves. Next, carbon dioxide absorbed by the leaves is converted into carbohydrates through photosynthesis to promote the growth of the tree.

Carbon dioxide is taken in through tiny openings in leaf surfaces called stomata. With the help of the sun, water and air are combined in the presence of chlorophyll to make sugars that provide energy to the growing tree. Some sugars are used in making new leaves, some in making new shoots, and the rest in making new wood. Some of the sugar moves to special locations in the tree or the roots, where it is stored for later use. Some are consumed through respiration. Sugar used in making new wood moves down a tree through the phloem. It is transported throughout the tree in the form of sap, a solution containing various sugars, water, growth regulators (hormones), and other substances. The term sap is also used to refer to the mineral-rich water that is taken up by roots and moved upwards through the outer portion of the xylem (Fig.2.5).

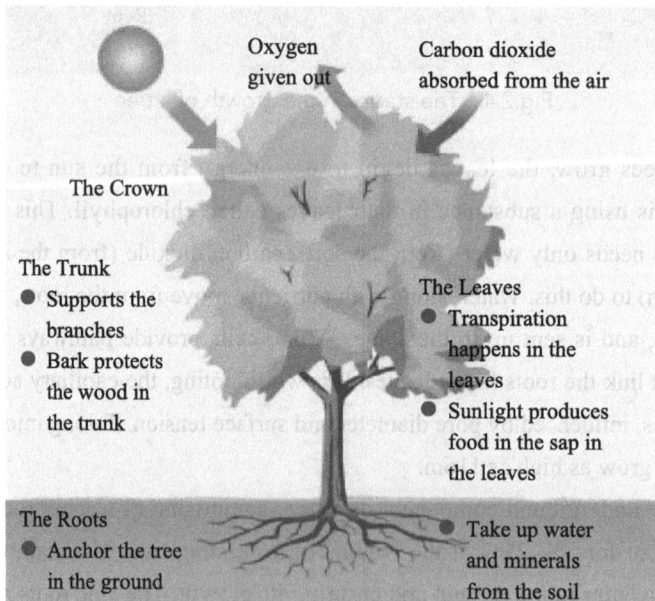

Fig.2.5　Standing tree

The extremely thin layer between the xylem and phloem produces new xylem inwards and phloem tissue outwards. This layer, called the cambium, completely sheaths the twigs, branches, trunk, and roots. This represents a season of growth resulting in a new continuous layer of wood throughout the tree (Fig.2.6).

Although sap moves down the tree through the phloem, the formation of the cambium requires nutrients. Accordingly, sap also needs to travel horizontally toward the center of the tree. Thus, wood rays provide for this horizontal movement, while also contributing to the storing of carbohydrates. Following periods of dormancy, rays serve as a channel of horizontal transport for stored materials to move from the center of the tree. Fig.2.7 illustrates the relative position of various portions of a tree trunk. Careful examination would help in gaining an understanding of the relationship between various layers of the tissue.

Fig.2.6 New growth occurs as a sheath covering the main trunk, branches, and twigs

Fig.2.7 Parts of a mature tree trunk

Tree growth refers to the process in which the shape and quality of trees increase continuously through cell division and expansion during growth and development. Trees are perennial plants, and its life cycle goes through young, mature, and overmature periods, until aging and death.

2.2.1 Height Growth in Trees

To begin the study of the development process, the growth of a young pine seedling will be considered. The seedling shown here has a well-developed root system and crown, which is typical of a 1~2 year old tree. With the beginnings of growth in early spring, buds at the tip of each branch swell while the tissue expands through the formation and growth of cells. The regions in which cells divide repeatedly to form new cells are called *meristematic regions*. Buds of similar appearances occur at the tip of each branch. The meristematic zone at the apex of the main trunk is of special significance because it controls, to some extent, the development of branches and shoots. It is called the *apical meristem*.

Cell division at the apical meristem serves to lengthen the main trunk. New cell production at this location is followed by cell elongation, resulting in *height growth*. As the trunk is built through the production of new cells during growth periods, the terminal bud moves upward, leaving new and expanding cells behind. Because trees grow from the apex rather than from the base, nails which are put into a tree at 2m above ground level will remain there regardless of the height to which the tree grows. This is called height growth.

The height growth process of trees shares similarities with the bricklaying work of workers. The position of the meristem at the top of the tree is equivalent to that of a bricklayer. As the wall is being built, the position of the bricklayer is constantly raised (Fig.2.8).

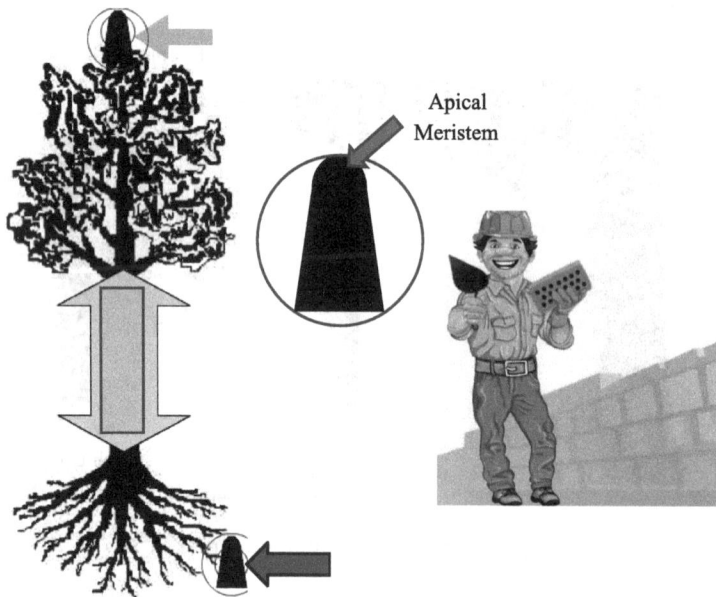

Apical Meristem

Fig.2.8　The height growth of a tree

2.2.2　Diameter Growth in Trees

The diameter growth of a tree results from meristematic activity in the cambium (lateral meristem). Inwardly, the primary cells of the cambium form *secondary xylem*. Outwardly, a *secondary phloem* is formed. As a result, the diameter of the trees increases continuously. The diameter growth of a tree caused by the differentiation of cambium is called secondary growth.

The tissues formed are called secondary tissues (Fig.2.9).

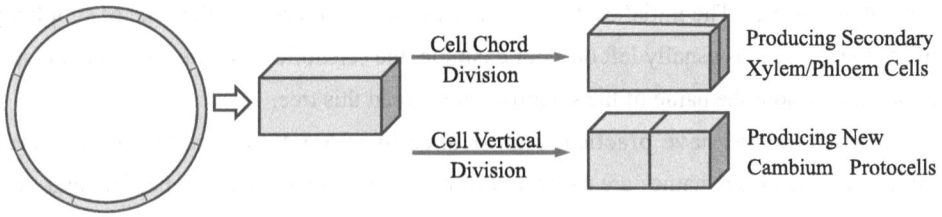

Fig.2.9 Diameter growth of a tree

The trunk is the main source of wood. It consists of four parts: the *bark, cambium*, *xylem* (*heartwood* and *sapwood*), and *pith* (Fig.2.10).

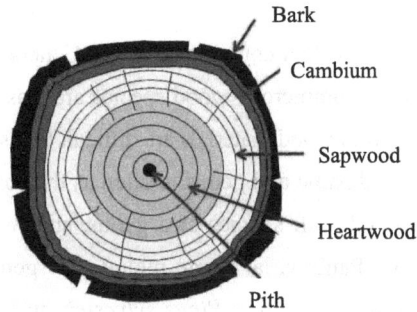

Fig.2.10 Parts of a tree in the cross-section

2.3 Nomenclature of Wood

Like other species, wood have their own names, such as pine, cypress and poplar. A type of wood can have different names in different places. Even it has several or more names in the same place. These different common names act as synonyms for a given tree species. Conversely, the tree species can be homonyms within foreign bodies. For example, the common name for *Zizyphus spinosus* in southern China refers to a species in *Aceraceae*. The same name refers to a species in *Rhamnaceae* in northern China. Additionally, the general concept of a pine also refers to a variety of woods in the Pinus species. At the same time, it also refers to the majority of coniferous wood, except cypress and fir. It is sometimes even used as a synonym for coniferous wood, resulting in a mixed usage of wood names. Therefore, it is necessary to know the scientific name of wood to better identify wood and promote timber trade.

2.3.1 Scientific Name

Scientific names are universally used and recognized. Scientific names are composed of Latin or other Latin languages, so they are also called Latin names.

Each scientific name includes the genus name and species name. The "double naming method" is adopted, and the species name is followed by the surname of the person who described the species. The initial of the generic name is capitalized. For the sake of simplicity, the name of the scientist is usually left out. For example, the scientific name for Masson pine is *Pinus massoniana* without the name of the scientist who named this tree.

Scientific names have practical significance in international exchange and scientific identification. Scientific names are namely scientific. But, due to language barriers and the large variety of wood species, it is not always easy to practically identify species with the naked eye. Therefore, the use of scientific name is very limited in the field of wood industry. Yet, tree species with few differences in macroscopic characteristics and materials have almost the same use value, thus in some cases it may be unnecessary to distinguish species.

2.3.2　Commercial Name

Wood used for market trading is called commodity wood. Generally, the classification of trees is based on the genus and material. Commercial wood species are classified into trees with similar macrostructures. The few differences in wood materials are difficult to distinguish by sight. Therefore, species shall be used in the standard name and genus is often as the commercial name for wood.

Some commercial wood includes all species of a genus, such as all species of Paulownia, which are traded under the name Paulownia. Some include the genus, such as pines (or hard pines), whose trade names are Pinus, including *Pinus sylvestris* and *Pinus tabulaeformis*. Some include species of a different genera, such as *Cyclobalanopsis glauca, Cyclobalanopsis glauca* and *Cyclobalanopsis glauca*.

2.3.3　Common Name

The common names for the types of wood are the name that are used locally to identify the wood. Due to variable naming habits and languages, different wood names are used. For example, so-called "beech" on the market actually refers to the wood of *Fagus spp.* in Fagaceae. However, actual beech is called *Zelkova spp.*

Therefore, the lack of standardized wood common names will inevitably bring about confusion names, causing obstacles towards the circulation of wood.

⫸ Exercises

1. How can you classify the vast number of plants found on this planet?

2. What is the source of wood? How does it come about?

3. How do a tree grow and develop its total volume?

4. How do you name wood?

Chapter ❸ »»»
Structure of Wood

Wood is a kind of natural materials. It is mainly composed of hollow, slender, spindle-shaped cells that are arranged along the growth direction of the trunk. When wood is obtained from trees, these cells and their arrangements can affect its strength, shrinkage, and texture. Thus, to study wood structures is of great importance to understand and identify wood.

Research on wood structures can be divided into three levels based on tools and the degrees of magnification:

- ◎ *Macroscopic structure*: characteristics observed by the naked eyes or by magnifiers.
- ◎ *Microscopic structure*: characteristics seen with the help of an ordinary optical microscope.
- ◎ *Ultrastructural structure*: characteristics of the wood cell wall revealed by X-ray and electron microscopy.

The investigation of wood structures plays an indispensable role in scientific research, rational processing, and value-added utilization of wood. This chapter mainly introduces the anatomical structures of wood at the macroscopic and microscopic levels.

3.1 Macroscopic Characteristics of Wood

The macroscopic characteristics of wood are worthy to research because they often give clues to the conditions under which wood was grown, provide an indication of its physical properties, and serve as an aid in wood identification. The macroscopic structure of wood is also known as the gross structure of wood, which refers to the structural characteristics of wood that can be observed with the naked eye or with a 10-times magnifying glass.

3.1.1 Three Directions of Wood

Fig.3.1 defines three directions of wood.

Longitudinal direction: the main axis direction of a trunk, which is also the direction along the grain of wood.

Radial direction: the normal direction of the growth ring (annual ring).

Tangential direction: the tangent direction of the growth ring (annual ring).

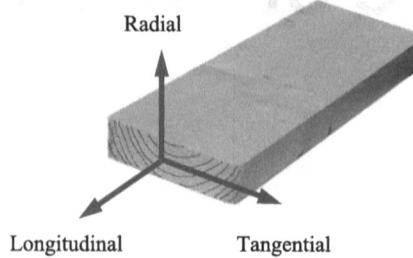

Fig.3.1　Three directions of wood

3.1.2　Three Distinct Surfaces of Wood

Illustrated in Fig.3.2 is a column-shaped piece of wood as it would appear if it is cut from a round cross-section. Notice that the macro features of the cross-sectional, radial, and tangential surfaces appear quite different. Wood differs not only in appearance depending upon the direction from which it is viewed, but in physical properties as well, which will be explained later.

Cross-sectional surface: the surface perpendicular to the longitudinal direction or the grain of the wood, is also known as the *end section*. The growth rings are concentric circles, which is the most important section used to identify wood. In application, owing to its high hardness and friction resistance, it can be used as chopping blocks or paving boards.

Radial surface: this surface can be obtained by cutting the wood through the pith and parallel to wood ray or perpendicular to the growth ring along the longitudinal direction of the trunk. In this section, the growth rings are parallel to each other and perpendicular to the wood rays. The lumber cut along the direction of the diameter has small shrinkage and is not easy to warp. Thus, it is suitable for flooring, wood rulers, and musical instruments.

Tangential surface: as shown in Fig.3.2(c), the knife is used to cut the wood perpendicular to the wood rays or parallel to the growth rings but not through the pith. The growth rings presents a parabola or an inverted "V" shape pattern on this section. The *figures* in this surface are very good-looking and suitable for furniture-making.

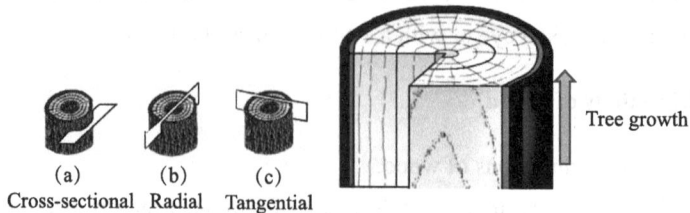

(a)　　　(b)　　　(c)
Cross-sectional　Radial　Tangential

Fig.3.2　Three distinct surfaces of wood

3.1.3　Growth Rings (Annual Rings)

The seasonal nature of growth was explained in Chapter 2. Growth in temperate zones is characterized as accelerating in early spring, and slowing in late summer before ceasing in autumn. For reasons explained in the next few paragraphs, this growth pattern usually results in the formation of one growth ring per year with distinct layers surrounding the ring. The trees grown in the temperate zone, frigid zone, and subtropics only grow one layer of wood per year. Thus, this ring is called the *annual ring* (Fig.3.3).

The number of annual rings is related to the properties of the wood. In terms of wood utilization, the physical and mechanical properties of wood are estimated by the number of annual rings one centimeter from the vertical growth ring on the cross section. Generally, the strength of coniferous wood is higher if the number of rings per cm is uniform. As for the broad-leaved wood, the strength of ring-porous wood species, is greater when the ring is wider.

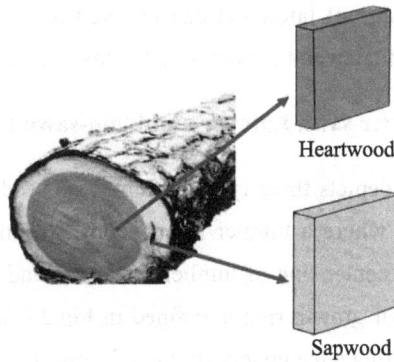

3.1.4　Heartwood and Sapwood

Examination of a wood cross-section often reveals a dark-colored circular area in the center, which is surrounded by a lighter-colored outer zone (Fig.3.4). The dark center area is known as *heartwood* and the lighter tissue as *sapwood*. It is important to note that heartwood is occasionally the same color as sapwood. It is in the sapwood where the living cells are found. The inner area of the sapwood, where most cells are dead, also serves to conduct water upwards in a living tree. However, heartwood no longer functions physiologically because all the cells in this section of the tree are dead.

Heartwood

Sapwood

Fig.3.3　Annual rings of wood　　　Fig.3.4　Heartwood and sapwood

3.1.5　Earlywood and Latewood

One growth or annual ring is made up of two parts (Fig.3.5). Near the pith, wood is formed at the early stages (spring or summer). This part is called *earlywood*. For trees growing in temperate, frigid, and subtropical zones with more rain and higher temperatures in spring and summer, water and nutrients are sufficient. Therefore, cell division speed in cambium is rapid, forming thin-walled and large-lumened cells. The earlywood of these cells has a low density and a light color.

On the side near the bark, wood in the same ring is formed at the late stage of growth. This part is called *latewood*. In autumn and winter, there is little rain and the temperatures are low. Accordingly, there is insufficient water and nutrients for tree growth. Thus, cambium cell division becomes slower and thick-walled and small-lumened cells occur. The wood of these cells has a high density and a deep color.

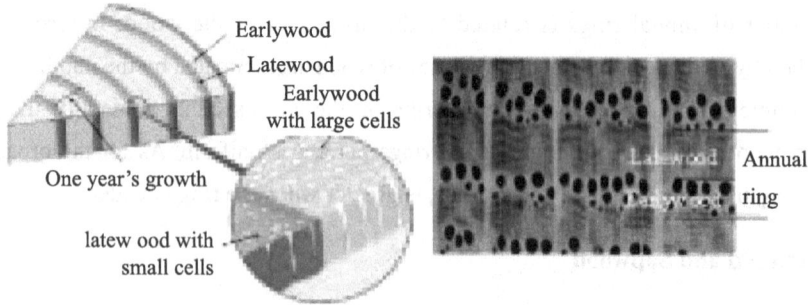

Fig.3.5　Earlywood and latewood

The percentage of latewood can be calculated using the following equation:

$$P = \frac{b}{a} \times 100\%$$

In which, *b* represents the latewood width within one growth ring (mm); a represents the total width of one growth ring (mm).

Percentage of latewood can be used to estimate the strength of wood. Generally, high percentage of latewood indicates high density and high strength.

3.1.6　Quarter-sawn Lumber and Plain-sawn Lumber

Fig.3.6 depicts three types of lumbers based on the location where a lumber is cut from. An angle between the center line of lumber thickness and the tangent line of growth ring is defined in Fig.3.6 with an aim to facilitate the understanding of these types of lumbers.

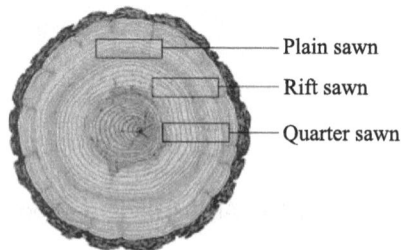

Fig.3.6　Types of lumbers

Plain-sawn lumber: when the angle is between 0° and 30°, the lumber is called a plain-sawn lumber or a flat-sawn lumber.

Rift-sawn lumber: when the angle is between 30° and 60°, the lumber is called a rift-sawn lumber.

Quarter-sawn lumber: when the angle is between 60° and 90°, the lumber is called a quarter-sawn lumber.

3.1.7 Pores in Hardwoods

Several differences exist between hardwood and softwood xylem, but the fundamental anatomical difference is that hardwoods contain specialized conducting cells called *vessel elements*. Normally, a number of vessel elements link end to end along the grain to form long tube-like structures known as *vessels*. The arrangement ensures that each branch of the crown receives water from many different roots, providing a safety feature against crown damage from the loss of one or more roots.

Because of their large diameter, vessels often appear as pores when viewed in the cross-section. In this view, they are often referred to as *pores*. Both the size and arrangement of pores can be used to classify hardwoods for purposes of identification.

Some hardwoods, for example, form large-diameter pores early in their growing season and much smaller pores later within a year (Fig.3.7); such woods are called *ring-porous*. Other hardwoods exhibit little variation in cell structure across a growth increment, thus forming rings that are difficult to detect. Because pores are about the same size throughout a growth ring, these woods are termed *diffuse-porous*. Note that *semi-ring-porous* wood, also called *semi-diffuse-porous* wood, refers to when the arrangement of pores is between ring-porous wood and diffuse-porous wood in a growth ring. Fig.3.7 shows the images showing ring- (ash), semi ring- (walnut), and diffuse-porous (birch) woods.

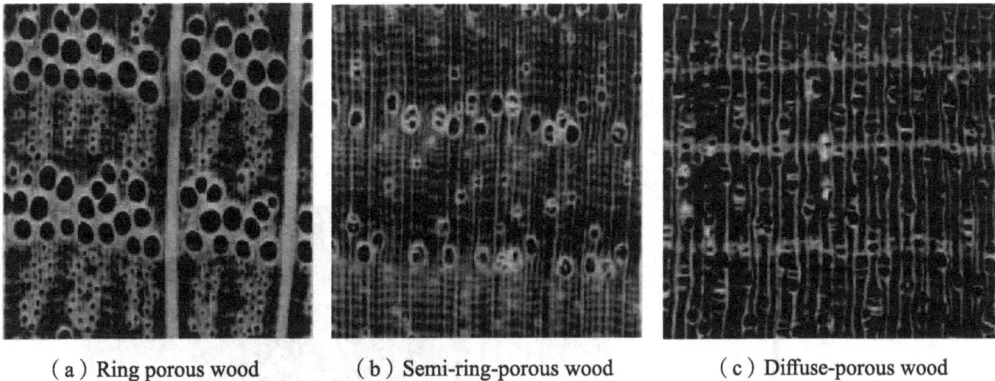

(a) Ring porous wood   (b) Semi-ring-porous wood   (c) Diffuse-porous wood

Fig.3.7   Pores in hardwoods

3.1.8 Wood Rays

Rays provide an avenue by which sap can travel horizontally either to or from the phloem layer. Virtually all woods contain rays. Wood rays represent radicalized lines with light color or glossiness in a cross-section (Fig.3.8). Many highly valued hardwoods, which are widely used for paneling, furniture, and in other decoration, are characterized by distinct ray patterns on radial and tangential surfaces. These are often of help in identification of wood species.

Fig.3.8　Wood rays

3.1.9　Intercellular Canal

The intercellular canal is a tubular intercellular space surrounded by secretory cells. The intercellular channel is called the *resin canal* in some softwoods that used to store resin, and is called *gum canal*, which is used to store gum in hardwoods. Six genera of *Pinaceae*, i.e. *Pinus*, *Larch*, *Spruce*, *Douglas-fir*, *Cathaya*, and *Keteleeria*, have normal resin canals. As shown in Fig.3.9, the softwood, *Pinus massoniana*, has normal resin canals.

Fig.3.9　Resin canal and resin products of *Pinus massoniana*

3.1.10　Wood Grain and Figure

The direction parallel to the long axis of most long-tapered fibers of wood is called the *grain direction*. Fibers are normally oriented as illustrated in Fig.3.10, with their length essentially parallel to the long axis of the stem, which is called the *straight grain*.

Trees in which fibers are spirally arranged around the trunk axis (Fig.3.10) are said to have a *spiral grain*. This condition is caused by anticlinal division where new cambial cell formation occurs in one direction only.

In some trees, the grain may spiral in one direction for several years and then reverse to the opposite direction (Fig.3.10). Wood produced in this way is said to have an *interlocked grain* (*reversing spiral grain*). The interlocked grain is evidently genetically controlled, very frequently occurring in some species and seldom, if at all, in others.

Fig.3.10 Wood grain

Wood figure refers to the beautiful patterns formed on the surface of the wood due to different growth rings, wood rays, parenchyma, knots, burl, texture, color, and sawing direction. Wood figure is closely related to wood structure. It helps identify wood and improve wood utilization value. The figure of softwood is relatively simple, while that of hardwood is rich and colorful.

3.2 Microscopic Characteristics of Wood

Gymnosperms evolved on earth before angiosperms, and conifers retain a relatively primitive cell structure compared with the more specialized and complex anatomy of hardwoods. As coniferous' anatomical characteristics are simpler, softwoods will first be discussed.

3.2.1 Microscopic Characteristics of Softwood

The anatomical structure of softwoods is relatively simple and features regular arrangements. These mainly include *longitudinal tracheids, wood rays, axial parenchyma cells*, and *resin canals*.

The great majority of softwood volume, 90%~95%, is composed of long, slender cells called *longitudinal tracheids*. Such cells are oriented parallel to the trunk axis. Longitudinal tracheids are about 100 times greater in length than in diameter, and are rectangular in cross-section (Fig.3.11). Tracheids have hollow centers (lumens) but are closed at the ends.

In one growth ring, earlywood has thin-walled cells with relatively large radial diameters and latewood has thicker-walled and smaller-diameter cells. Again, referring to Fig.3.11, most of the pits that mark the radial cell walls are of the bordered type. Such pits characterize tracheid-to-tracheid linkages. Hence, the location of a pit is usually matched with another pit in another adjacent longitudinal tracheid. The rows of small, lemon-drop-shaped pits mark the points at which ray parenchyma cells contact the longitudinal tracheids.

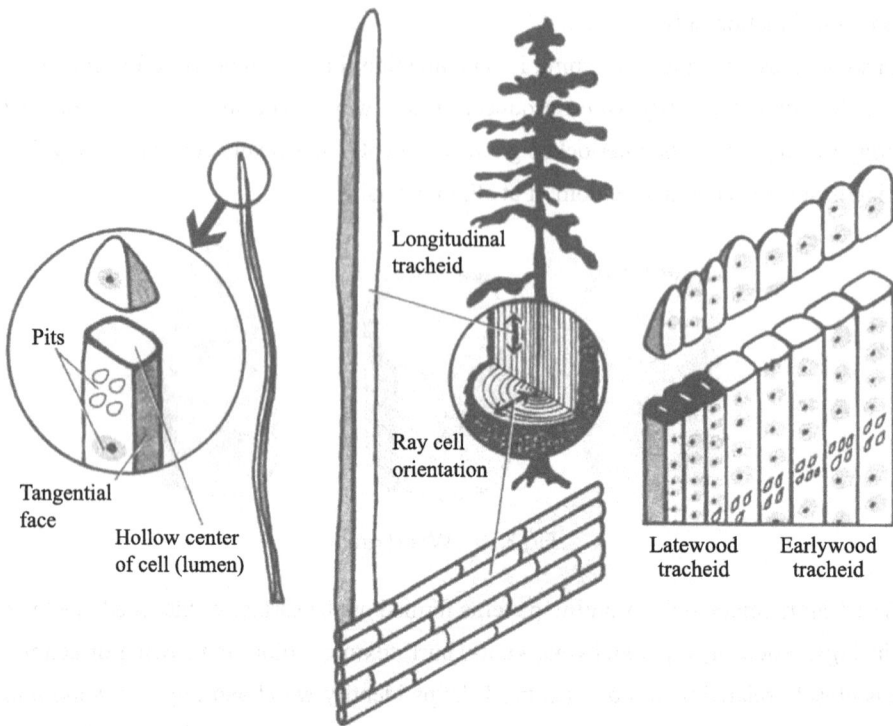

Fig.3.11 Arrangement of cells in softwood

Rays are important identification features in conifers when viewed microscopically. As such, magnification produces valuable information on the height or width of ray, types of cells composing the rays, and types of pitting or sculpting in the cell walls. Most wood rays are composed of ray parenchyma cells, which store nutrients in the sapwood and transmit water and nutrients. In heartwood, ray parenchyma cells are dead. Uniformly narrow rays characterize softwoods, except where horizontal resin canals are present. Viewed tangentially, softwood rays are from one-to-many cells in height but are usually only one cell wide (*uniseriate*). A few species-such as *redwood* commonly have rays that are two cells wide, at least at some points along the rays. These rays are called *biseriate rays*. In general, species that have normal longitudinal resin canals have horizontal resin canals. These small canals occurring in the middle of specialized rays are called *fusiform rays*. Some species, such as pine, *spruce, larch, cedar, hemlock*, and *Douglas fir*, have thick-walled cells in their rays, called ray tracheids. Ray tracheids are similar to those existing in longitudinal tracheids, because both have *bordered pits*.

A small portion of the volume of some softwoods is composed of *axial parenchyma cells*. When mature, these cells have the same general shape as longitudinal tracheids, although they often subdivide a number of times, along their length prior to forming secondary walls. These results in a mature parenchyma that usually appears as longitudinal strands of short cells butted end-to-end in series. The thin-walled and *simple-pitted* parenchyma account for as much as 1% or

2% of the volume of some softwoods. The presence or absence of longitudinal parenchyma cells helps in identification of softwood species. The genera *Pinus* and *Picea* don't have *longitudinal parenchyma*.

Resin canals are tubular passages in wood that are intercellular spaces surrounded by special parenchyma cells called *epithelial cells*. These epithelial cells exude pitch or resin into the canals. In turn, serving a protective function for the tree by exuding pitch to seal off wounds caused by mechanical damage or boring insects. Resin canals are present in all species of four genera within the family *Pinaceae*: pines (*Pinus* spp.), *spruces* (*Picea spp.*), *larches* (*Larix spp.*) and *Douglas-fir* (*Pseudotsuga menziesii*). The production of resin canals in response to injury or other traumatic events is not restricted to those genera that produce resin canals of the normal type. Although the tendency to produce traumatic canals is greater in some woods than in others, resin canals of this type may occasionally occur in almost any type of softwood. *Hemlock* (*Tsuga*), *redwood* (*Sequoia*), and *true fir* (*Abies*) are examples of wood that do not have canals of the normal type but commonly form resin canals in response to injury.

Fig.3.12 shows the microscopic anatomical structure of *masson pine*. Examination of the transverse surface reveals numerous longitudinal tracheids in the cross-section. Latewood tracheids of one annual growth layer lie to the right. In earlywood, these tracheids appear on the immediate left side. Transition is relatively abrupt. Cells in earlywood are thin-walled and large-lumened. In latewood, they are thick-walled and small-lumened. A resin canal is surrounded by short, thin-walled epithelial cells. Thin-walled longitudinal parenchyma lie to the outside of the epithelial cells. Viewed transversely, the transversely sectioned row of ray, tracheids and the size of wood ray can be identified. In the radial surface, numerous conical-shaped bordered pits and the radial surface of longitudinal tracheids can be observed. They mark locations of matching pits in adjacent rows of tracheids. Earlywood tracheids are bluntly-tapered in this view, whereas end walls of the narrower latewood are more angular. A longitudinally-sectioned uniseriate and

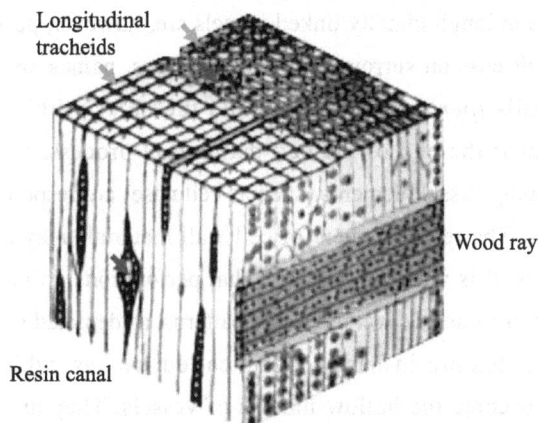

Fig.3.12 Three-dimensional representation of distinct-ring softwood

heterogeneous ray composes of both ray tracheids and ray parenchyma. To the extreme left of the tangential surface, a septate longitudinal parenchyma cell is visible. This is adjacent to a large longitudinal resin canal, which is surrounded by epithelium. Longitudinal tracheids appear sharply tapered tangentially, rather than rounded as in a radial view.

3.2.2 Microscopic Characteristics of Hardwood

Hardwood contains vessels with the exception of a few species. Hardwood is composed of at least four major kinds of cells, including *vessel elements, fiber tracheid, axial parenchyma cells*, and *ray parenchymas*. The classification and functions of cells in hardwood is shown in Fig.3.13.

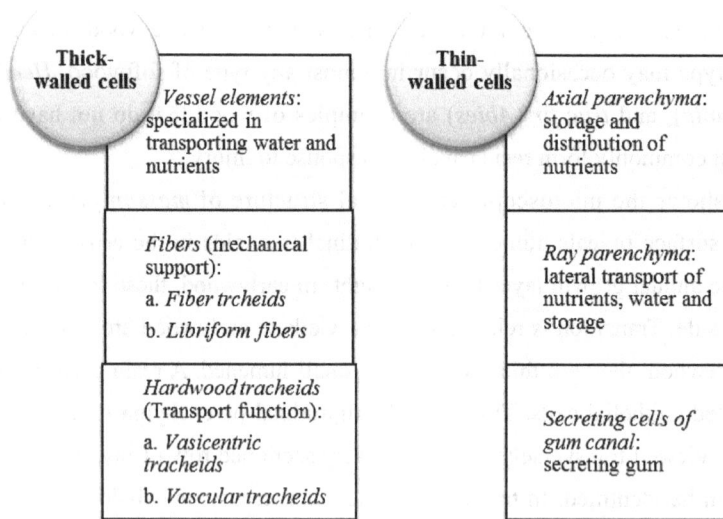

Thick-walled cells

Vessel elements: specialized in transporting water and nutrients

Fibers (mechanical support):
a. *Fiber trcheids*
b. *Libriform fibers*

Hardwood tracheids (Transport function):
a. *Vasicentric tracheids*
b. *Vascular tracheids*

Thin-walled cells

Axial parenchyma: storage and distribution of nutrients

Ray parenchyma: lateral transport of nutrients, water and storage

Secreting cells of gum canal: secreting gum

Fig.3.13　Major types of cells in hardwood

Vessels are uniquely suited to serve as channels for fluid conduction. The relatively membrane-like vessels, divide pit pairs and connect other cells, such as fiber tracheids, end-to-end. Common end walls in longitudinally linked vessels are, however, perforated by unrestricted holes. To facilitate the discussion surrounding these features, names are given to the common vessel element end walls (*perforation plates*) and the holes within them (*perforations*). Perforations develop near the end of the cell maturation process. Certain enzymes in the protoplast of the developing vessel elements (such as cellulase) are responsible for the dissolving of the perforation plates. The rearrangement of cell wall materials may also be involved in the formation of perforations. It is interesting to note that perforations do not develop in a random fashion. Instead, they follow one of several definite patterns, as depicted in Fig.3.14.

Another important feature in hardwood is called *tyloses*, which are outgrowths of parenchyma cells that become the hollow lumens of vessels. They are commonly formed in hardwood as a result of wounding and the prevention of water loss around the damaged area.

Tyloses also form in a number of species during the transition from sapwood to heartwood. They may also develop as a result of infection from fungi or bacteria.

Simple   Scalariform   Foraminate

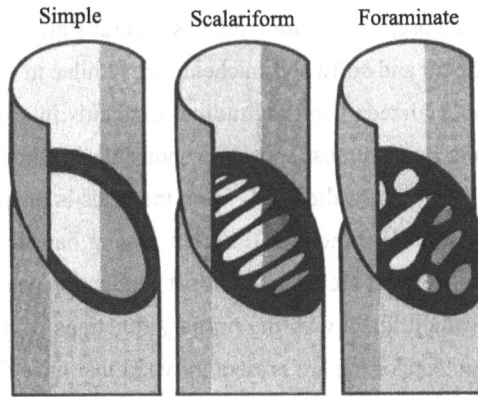

Fig.3.14   Types of perforation plates end-to-end connecting vessels

Prior to tylosis formation, enzymatic action partially destroys the membranes among vessel-to-parenchyma pit pairs. At the same time, the cytoplasm of the parenchyma cells begins to expand. Accordingly, parenchyma cell membranes protrude through the pit pairs into the vessel lumen. This protrusion is called a *tylosis*. Tyloses often partially or completely block the vessels in which they occur, resulting in either detrimental or beneficial consequences based on the use of the wood. The obvious existence of tyloses in heartwood of white oak, and the relative lack in red oak, justify why white oak is preferred in the manufacture of barrels, casks, and tanks for the storage of liquids (Fig.3.15). More specifically, white oak's heartwood, with its tightly plugged vessels, is used almost universally to manufacture barrels for whiskey, scotch, bourbon, and wine. On the contrary, the red oak is generally avoided being used for making barrels. In addition, wood in which tyloses are well-developed may be difficult to dry or impregnate with decay-preventive or stabilizing chemicals.

Vasicentric tracheids and *vascular tracheids* are the two tracheids that may appear in some types of hardwood. However, they are rarely seen and are not dominant.

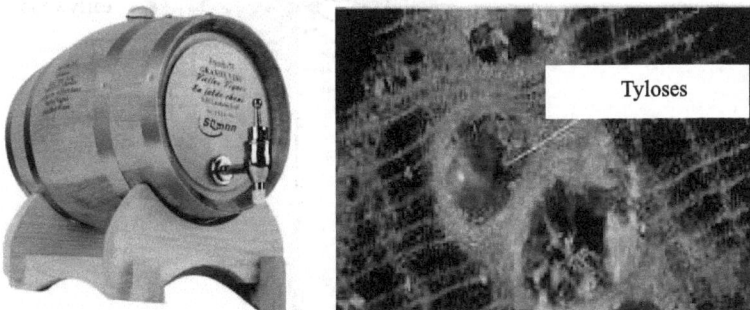

Tyloses

Fig.3.15   White oak with rich tyloses

In the context of wood morphology, the term *fiber* refers to a specific cell type. Fibers, or fiber tracheids as they are more properly called, are long, tapered, and usually have thick-walled cells of hardwood xylem. A simple observation demonstrates great similarities to the longitudinal tracheids in softwood. But a closer examination reveals several significant differences.

Although hardwood fibers and softwood tracheids are similar in shape, the function of their respective fiber is more specialized. The longitudinal tracheids in softwoods serve as primary avenues of conduction while being almost entirely responsible for the strength of the wood. The presence of a high proportion of thin-walled earlywood tracheids is invariably related to low wood strength. However, the situation is somewhat different among hardwoods, where both kinds of longitudinal fibers and vessel element are common. Most conduction occurs through the specialized vessels, leaving the thick-walled fibers with the primary functions of mechanical support. Fibers that are highly specialized in supporting are present in wood that have the most specialized vessel members. The density, and thus strength, of hardwood is therefore generally related to the portion of wood volume occupied by fibers relative to that accounted for by vessels. As a general rule, the higher the proportion of thick-walled fibers, the higher the strength of the wood.

The walls of fiber tracheids are marked by pits of the bordered type. Fiber-to-fiber pit pairs are normally bordered, whereas fiber-to-parenchyma pitting is typically half-bordered. A variation of fiber, known as a *libriform fiber*, is marked by simple pits, rather than bordered ones. Libriform fibers occur in considerable numbers in some species. Fibers and vessels are seldom connected by pit pairs.

Parenchyma cells are thin-walled storage units (Fig.3.16). In hardwood, such cells occur in the form of long, longitudinally-tapered cells, short, brick-shaped epithelium around gum canals (in some species), and ray cells. During the process of cell maturation, the longitudinal form of parenchyma is often divided into a number of smaller cells through the formation of cross-walls. Parenchyma arrangements are described based on the relationships between parenchyma cells and pores: if the

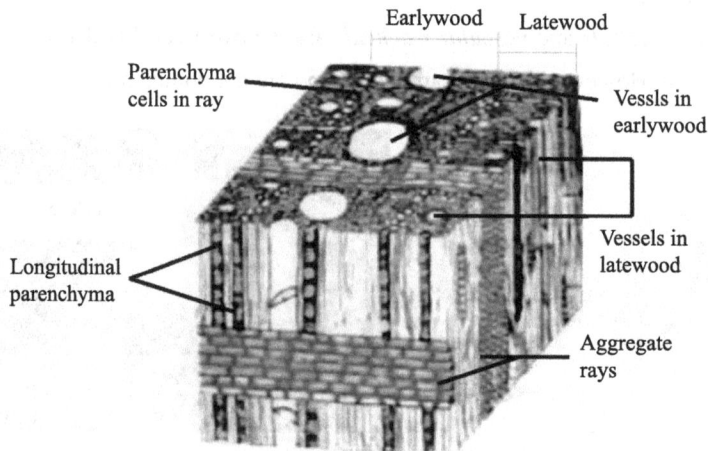

Fig.3.16　Microscopic anatomical structure of hardwoods

parenchyma cells make direct contact with the pores, they are *paratracheal parenchyma*; however, if they are separated from the pores by rays or fibers, they are *apotracheal parenchyma*.

The tangential range of hardwood rays measures from 1 to 30 or more cells in width. In comparison, softwood rays are generally one or, rarely, two cells in width. Also, unlike softwoods, the cells of hardwood rays are all of the parenchyma type.

3.3 Wood Identification

Wood recognition is based on the structural characteristics to identify the species of wood. Traditional recognition methods mainly identify wood species according to the macro and micro structural characteristics of wood.

3.3.1 Identification Based on Macroscopic Features of Wood

The standard magnifier used for the macroscopic identification of wood is the 10-power (10x) lens. The examiner should hold the lens close to the eye under good lighting, then move the piece of wood towards the lens until it comes into focus (Fig.3.17). By butting hand-to-hand and hand to cheek, the eye, the lens and the wood sample can be aligned in a constant position with maximum visibility of the cell structure. Generally, the end-grain surface of the wood sample is polished with a sharp knife and moistened with water. When practicing identification of wood species, an acquaintance with the pertinent structural features, both gross and macroscopic, should first be acquired.

Heartwood and sapwood, growth rings, vessels, rays, and parenchymas need to be observed at the cross-section surface. Combining wood color, luster, texture, structure, figure, odor and taste, weight and hardness of wood, wood can be comprehensively judged and identified.

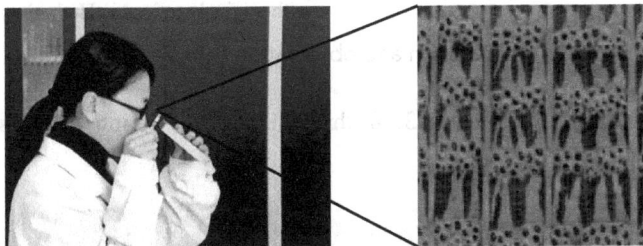

Fig.3.17 The right use of a lens and cross surface of oak

3.3.2 Identification Based on Microscopic Features of Wood

Since wood is an anisotropic natural material, observation of three surfaces of wood sample, namely cross surface, radial surface, and tangential surface greatly helps understand its anatomical structure at the cellular level. For observation purposes, wood slices shall be prepared by either cutting by hand or with a slicing machine.

Freehand slicing: using a sharp knife (art knife or blade), gently drag over the surface of a wood sample and cut into small and thin slices.

Machine slicing: small cubic wood samples of three standard surface should be firstly made to 20mm or more in length, width, and height. After being boiled with water or softened by chemicals, these samples of different surfaces are cut to a thickness of 10~20μm on a slicer. Next, optical microsections can be undertaken by dyeing, dehydration, transparence and sealing, and so on. These microsections are set under an optical microscope to observe their microscopic anatomical characteristics, that is, the morphology and arrangement characteristics of various kinds of cells and tissues.

（a）Block cutting　　　　（b）Cutting suitable sample　　　　（c）Wood slicing

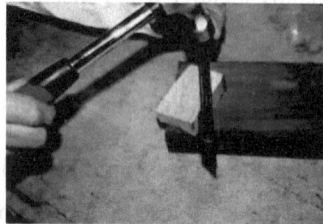

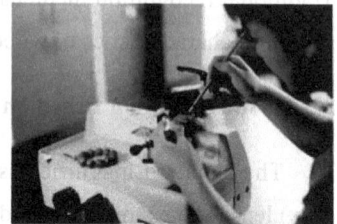

（d）Observation by optical microscope　　（e）Comparison of existing sections
with original sections in Herbarium

Fig.3.18　Preparation and observation of wood microslides

Based on the observations in Fig.3.18, three distinctive section surfaces of wood can be obtained as shown in Fig.3.19.

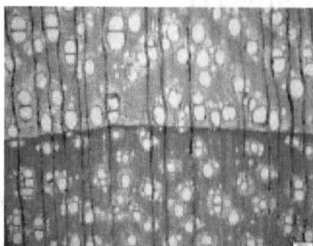

（a）Cross sectional surface　　　　（b）Radial surface　　　　（c）Tangential surface

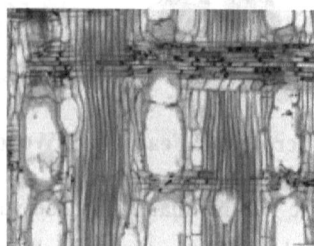

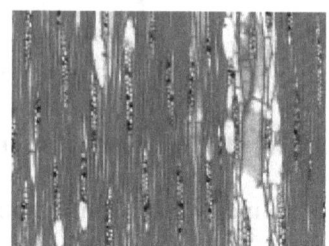

Fig.3.19　Anatomical structures of Cinnamomum camphora in three sections under a microscope

Because the microscopic identification involves many anatomical features of wood, it greatly improves the accuracy of identification of wood species. However, the process of this method is relatively complex.

3.4 Introduction of Main Commercial Woods

3.4.1 Domestic Woods

(1) Chinese Fir

Cunninghamia lanceolata (Lamb.) Hook. is a unique fast-growing plantation tree species in China. The wood is yellowish white, with the heartwood sometimes reddish brown. It is soft, delicate, fragrant, straight grain, and easy to process. Its specific gravity is 0.38. It has a strong corrosion resistance and is susceptible to termites. Chinese fir can emit a special aroma. It has the advantages of a straight shape, uniform structure, light weight and toughness, moth resistance, and corrosion resistance. Thus, it is widely used in construction, bridges, doors and windows, furniture, decoration, wooden crafts, and so on (Fig.3.20).

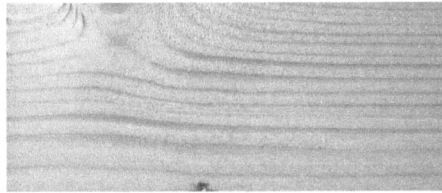

Fig.3.20   Chinese fir

(2) Chinese Red Pine

Pinus massoniana Lamb. also names green pine or mountain pine. It has a light yellowish-brown color, straight grain, coarse structure, specific gravity of 0.39~0.49, rich resin, and weak corrosion resistance (Fig.3.21). It is used as raw materials for construction, sleeper, pillar, furniture, and within the wood fiber industry (rayon pulp and papermaking).

Fig.3.21   Chinese red pine

(3) Larch

Larix gmelinii (Rupr.) Kuzen. is the main tree species found among in the coniferous forests of Daxing'an Mountains in China. It has resin, a straight grain, fine structure, and a specific gravity of 0.32~0.52 (Fig.3.22). It can be used for building construction, civil engineering, pole, boat cart, fine wood processing, and wood fiber industrial raw materials.

Fig.3.22　Larch

(4) Korean Pine

Pinus koraiensis Sieb. et Zucc. is an evergreen tree. It is produced in the Changbai Mountain Area of Northeast China. The wood is light, soft and fine. It has a straight gain and a strong corrosion resistance (Fig.3.23). It is an excellent form of timber for buildings, bridges, sleepers, and furniture.

Fig.3.23　Korean pine

(5) Chinese Cypress

Cupressus funebris Endl. is fine in texture and has strong quality and water resistance. It is commonly used in temples, palaces and courtyards. The wood is a fat wood made of fine materials, straight lines, fine structure, and corrosion resistance (Fig.3.24). It is used for construction, vehicles, bridges, furniture, and appliances.

Fig.3.24　Chinese cypress

(6) Mongolian Scotch Pine

Pinus sylvestris var. *mongolica* Litv. is one of the main fast-growing tree species in Northeast China. The wood is strong and has a straight grain (Fig.3.25). It can be used for construction, furniture and other materials.

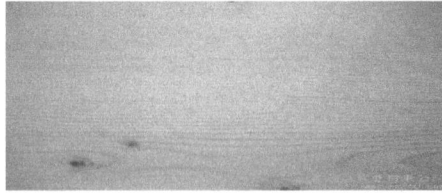

Fig.3.25 Mongolian scotch pine

(7) Chinese Spruce

Picea asperata Mast. is a unique tree species in China. The wood has few knots with a yellowish white color (Fig.3.26). It is light and soft, straight in grain, fine in structure and easy to process. Its specific gravity is 0.55~0.66. It has elastic properties and good resonance performance. It can be used for construction, aircraft, musical instruments (piano, violin), vehicles, furniture, appliances, boxes, planed plywood and veneer, as well as wood fiber industrial raw materials.

Fig.3.26 Chinese spruce

(8) Camphorwood

Cinnamomum camphora (L.) Presl., as a traditional precious tree species in China, has many advantages. These include its beautiful figure, hard wood, strong fragrance, decay resistance, and insect resistance (Fig.3.27). It is an ideal material for advanced architecture, shipbuilding, furniture, and sculpture.

Fig.3.27 Camphorwood

(9) Ash

Fraxinus mandshurica Rupr., can be found in many zones in China. It grows quickly in most soils, yet it is currently planted in forestry plantations of better quality soils. In the winter, the buds on the trees are black in color. Furthermore, ash has a very distinct leaf. Ash wood has a light-yellow color with a distinct grain pattern (Fig.3.28). It is a dense hardwood with good elastic properties.

Fig.3.28　Ash

(10) Oak

Xylosma racemosum (Sieb. et Zucc.) Miq. has the characteristics of a hard material. It has a large strength-weight ratio (air-dry density 0.63~0.72g/cm^3), beautiful figure, anti-corrosion properties and water humidity resistance (Fig.3.29).

Fig.3.29　Oak

(11) Walnut

Juglans mandshurica Maxim. Narrow sapwoods tend to be white. Of these, walnut remains a favorite for furniture, paneling, musical instruments, turned bowls, relief carvings, and sculpture. Furthermore, walnut's multiple advantages-shock resistance, strength, and stability-make it perfect for shotgun and rifle stocks (Fig.3.30).

Fig.3.30　Walnut

(12) Cork

Phellodendon amurense Rupr. is hard in nature, resistant to water and moisture, strong in rot resistance, and beautiful in figure. Additionally, it is lustrous, soft, and easy to process. It can be used for gunstock, aircraft, shipbuilding, construction, plywood, and furniture (Fig.3.31).

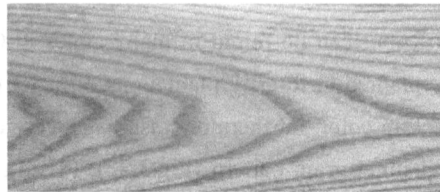

Fig.3.31　Cork

(13) Oriental White Oak

Quercus glauca (Thunb.) Oerst. This kind of wood is tough and can be used for piles, vehicles, ships, tool handles, etc (Fig.3.32).

Fig.3.32 Oriental white oak

(14) Fast-growing Woods

Fast-growing tree species refer to those experiencing fast growth, early maturity, and short rotation period. *Robinia pseudoacacia*, poplar, eucalyptus, and willow are common fast-growing species in China. The wood made from these fast-growing trees is called fast-growing wood. It is usually soft in texture, and is generally used as raw materials for paper-making, construction and wood-based panels. Some of them are used to make furniture. However, these wood materials must be treated prior to use.

3.4.2 Imported Woods

(1) Oak

Oak trees grow slowly. Much of the oak woodland was cut down years ago and used for construction, shipbuilding and for making barrels. Oak can live for hundreds of years. In open areas it develops a large central trunk and a broad crown. There are many varieties in the oak family. These can be divided into white oak and red oak. Red oak is mainly produced in North America, Europe and Turkey. Meanwhile, white oak is mainly produced in Asia, Europe and North America.

When used in furniture, flooring and other interior decoration, white oak and red oak do not need to be distinguished. However, for sealed barrel products, only white oak can be used, as white oak heartwoods contain abundant thylosed with an impermeability superior to red oak (Fig.3.33).

(2) Beech

Beech is a common tree in Europe, and it grows best in limestone soils at generally slow rates. It has a smooth grey bark and only bears seeds after 40 years. These seeds are contained in a triangular shell called a mast.

Beech wood is reddish or light brown in color (Fig.3.34). It is a close-grain wood, thus it is durable and hard-wearing. Beech is readily available and very popular.

(3) Maple

Maple is the national tree of Canada. These trees are a hardwood and of medium height. They are related to the sycamore tree. The leaves are very distinct and there are a number of varieties with different-colored leaves. For example, one kind of maple tree that is very common in Irish gardens has wine-colored leaves (Fig.3.35). Much of the maple used in China is imported.

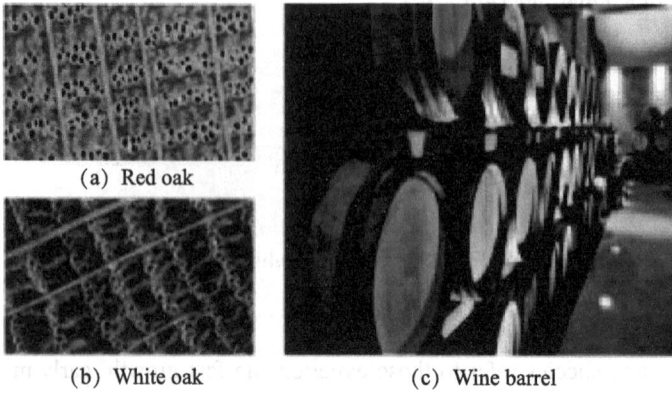

(a) Red oak

(b) White oak

(c) Wine barrel

Fig.3.33　Differences and white oak wine barrel

Fig.3.34　Beech tree, leaves, seed (mast) and wood

Fig.3.35　Maple

(4) Douglas Fir

Douglas fir is a native tree of North America. It is sometimes called Oregon pine or Colombian pine. It grows quickly and is tall. It likes well-drained soils and sheltered areas away from high winds. The needles grow all around the twig, which has reddish-brown buds. The pine cones are oval in shape, making the tree easy to identify.

The wood is a reddish-brown color (Fig.3.36). The grain is decorative, and the annual rings are clearly visible. It is a tough and durable wood that will withstand heavy wear. Although the wood is light, it is very strong. It is resistant to decay and acid attacks. It is mainly used in structural panels, gates and outdoor furniture, plywood, and railway sleeper.

Fig.3.36　Douglas fir tree, needles, cone and wood

(5) Southern Pine

Southern pine, also known as southern yellow pine, is a group name of four tree species, including longleaf pine (*Pinus palustris*), shortleaf pine (*Pinus banksiana* Lamb), slash pine (*Pinus ellitti*), and loblolly pine (*Pinus taeda*). These grow in the vast area of southern United States.

Southern pine not only has a beautiful natural appearance, but also has the strongest and toughest wood among coniferous trees (Fig.3.37). It is not easily damaged by collision and is extremely wear-resistant with good nail-holding abilities. Due to its excellent performance, southern pine is widely used in various types of buildings under various conditions, such as outdoor landscape facilities, and is known as "the world's top structural wood". Because of its natural porous cellular structure, preservatives can penetrate the inner layer of wood evenly and remain it for a long time to resist the invasion of various molds, termites, and other microorganisms.

Fig.3.37　Longleaf pine

(6) Radiata Pine

Latin name: *Pinus radiata* D. Don. Radiata pine is a high-quality softwood with a medium density, uniform structure, average shrinkage efficiency, and strong stability. These intact logs are free from decay, heart rot and insect bites. Their wood material has a good nail-holding ability, strong permeability, and is easy to be antiseptic, dry, solidified, and colored (Fig.3.38). Radial pine is widely used in construction, plywood, fiberboard, particleboard, pulp, furniture, poles, fences, and sleepers.

Fig.3.38　Radiata pine

(7) Eastern Hemlock

Hemlock is moderately hard yet moderately light (average specific gravity 0.47). Once buffed, their coloring is pale brown, sometimes with a faint reddish or purplish tinge (Fig.3.39). Its sapwood is always distinct from heartwood. It has a medium texture and fairly uneven grain. Its latewood is distinctly dense, occupying one-third or less of the ring, transition variable, semi-abrupt to abrupt. The normal resin canals are absent. Its rays are very fine, which are not visible to the naked eye.

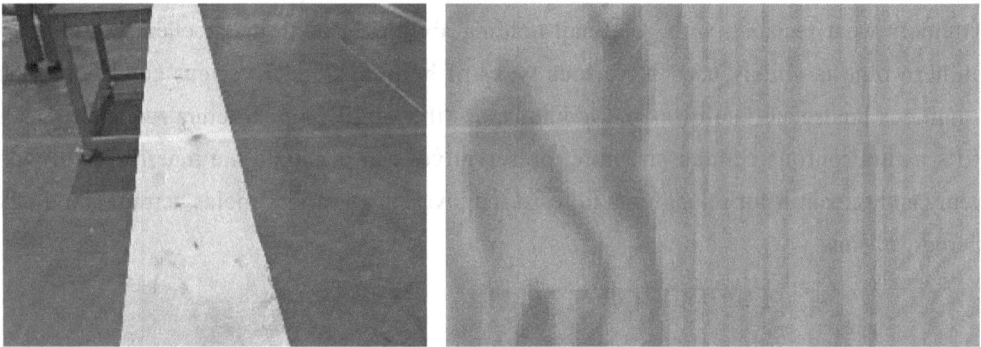

Fig.3.39　Eastern hemlock

(8) North American SPF

White spruce, alpine fir, and lodgepole pine are the main coniferous species in the S-P-F (spruce-fir-pine) combination (Fig.3.40). They have the same origins and many shared characteristics. The kiln-dry SPF sawn lumber is mainly used for the structural framework of various types of civil, commercial, industrial, and agricultural buildings. It is also widely used in the manufacture of prefabricated houses, wooden truss roofs and other structural components, as well as for terraces, roof panels, and decorative parts. In addition to producing high-quality structural materials, SPF can also be used to make very attractive and economical solid wood furniture.

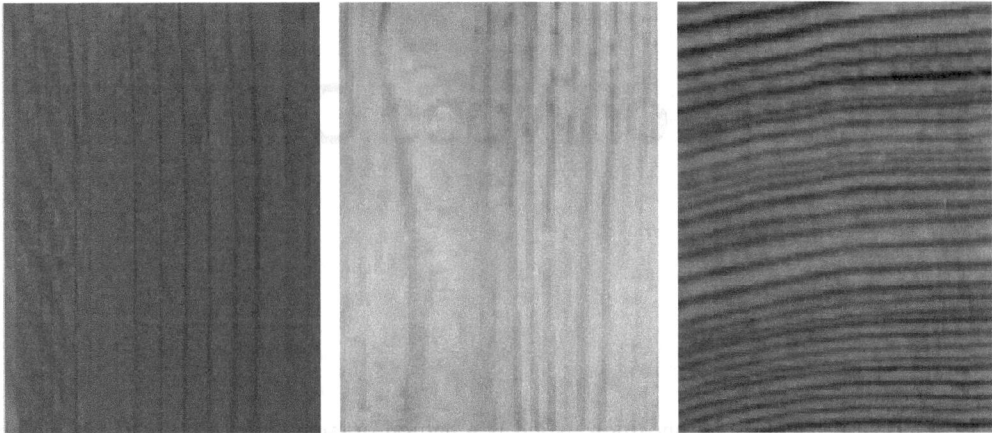

Fig.3.40   White spruce, alpine fir and lodgepole pine

Exercises

1. Summarize the differences in the macroscopic anatomical structures between hardwoods and softwoods.

2. Summarize the differences in the microscopic anatomical structures between hardwoods and softwoods.

3. Briefly introduce the anatomical characteristics of at least three domestic commercial woods and three imported commercial woods.

Chapter ❹ »»»
Structure of Wood Cell Wall

In terms of chemical composition, wood is a natural organic composite, which is mainly composed of cellulose, hemicellulose, lignin, and wood extractives. The character of wood depends on its chemical composition and structure. Understanding the chemical composition and structural characteristics of wood is helpful towards the understanding and utilization of wood.

4.1   Chemical Components of Wood Cell Wall

Wood is composed principally of carbon, hydrogen, and oxygen. Tab.4.1 details the chemical composition of a typical North American wood. It shows that carbon is the dominant element on a weight basis. Furthermore, wood contains inorganic compounds that still exist after high-temperature combustion in abundant oxygen. Such residues are known as ash. Ash can be traced back to incombustible compounds, which contain elements such as calcium, potassium, magnesium, manganese, and silicon. As domestic wood has a very low ash content, particularly a low silica content, it is very important from the standpoint of wood utilization. Wood that typically has a silica content of greater than 0.3% (on a dry weight basis) will make good cutting tools dull. Silica contents exceeding 0.5% are relatively common in tropical hardwoods. In some species, it may exceed 2% in weight.

Tab.4.1   Elemental composition of wood

Element	Percentage	Element	Percentage
Carbon	49%	Nitrogen	0.1%
Hydrogen	6%	Ash[*]	0.2%~0.5%
Oxygen	44%		

*As high as 3.0%~3.5% in some tropical species.

As shown in Fig.4.1, the chemical components of wood can be divided into primary and secondary components based on their content and function in wood. The main chemical components of wood are cellulose, hemicellulose, and lignin, which are the main substances that form cell walls in wood. The secondary components are extract and ash, which serve as internal substances in cell lumen, with a small amount found in cell walls.

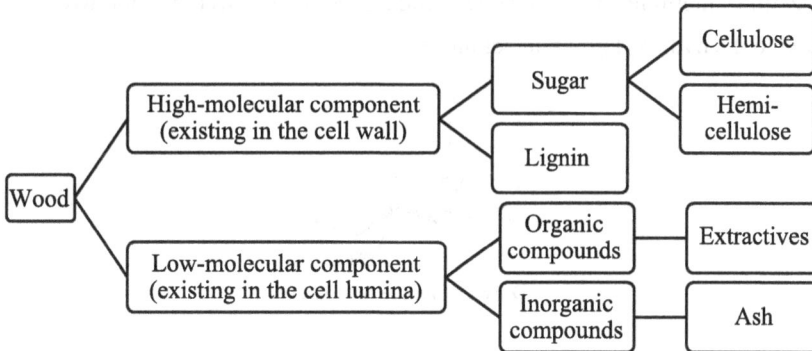

Fig.4.1   Molecular composition of wood

As for the main components of cell walls-cellulose, hemicellulose, and lignin, their respective proportions in softwoods and hardwoods are different (Tab.4.2).

Tab.4.2   Content comparison of three main components in softwoods and hardwoods

Species	Cellulose	Hemi-cellulose	Lignin
Softwoods	(42 ± 2) %	(27 ± 2) %	(28 ± 3) %
Hardwoods	(45 ± 2) %	(30 ± 5) %	(20 ± 4) %

4.1.1   Cellulose

(1) Chemical Structure of Cellulose

Cellulose is synthesized within living cells from a glucose-based sugar nucleotide. Cellulose $(C_6H_{10}O_5)_n$ (n is the polymerization degree) has a degree of polymerization, n, which may be as large as 10,000. The structural relationship between glucose and cellulose is depicted in Fig.4.2. Cellulose is a linear polymer composed of β-D-glucopyranoses linked with β-1-4-glycoside and C_1 chair conformations, while two glucose groups condense to form cellobiose.

A cellulose molecule is made of up to 10,000 glucose units, for which a very large structure might be envisioned. Although large from a molecular viewpoint, the longest cellulose molecules are about 10μm (1/1000cm) in length and about 8 Å in diameter (1Å = 1/100,000,000cm), which is too small to be seen even with an electron microscope.

As commonly proven among research, the chemical structure of cellulose (Fig.4.2) has some unique characteristics. In addition to the glucose groups at both ends, each glucose group has three free hydroxyl groups, which are located at positions C_2, C_3 and C_6. Therefore, a cellulose molecule can be expressed as $[C_6H_7O_2(OH)_3]_n$, in which secondary alcohol hydroxyl groups are located at positions C_2 and C_3, and a primary alcohol hydroxyl group is located at position C_6. Their chemical reactivity is different, which has an important influence on the properties of cellulose. These properties include esterification, etherification, oxidation and graft copolymerization, intermolecular hydrogen bonding, cellulose swelling, and hydrolysis, which are all related to the hydroxyl group in cellulose.

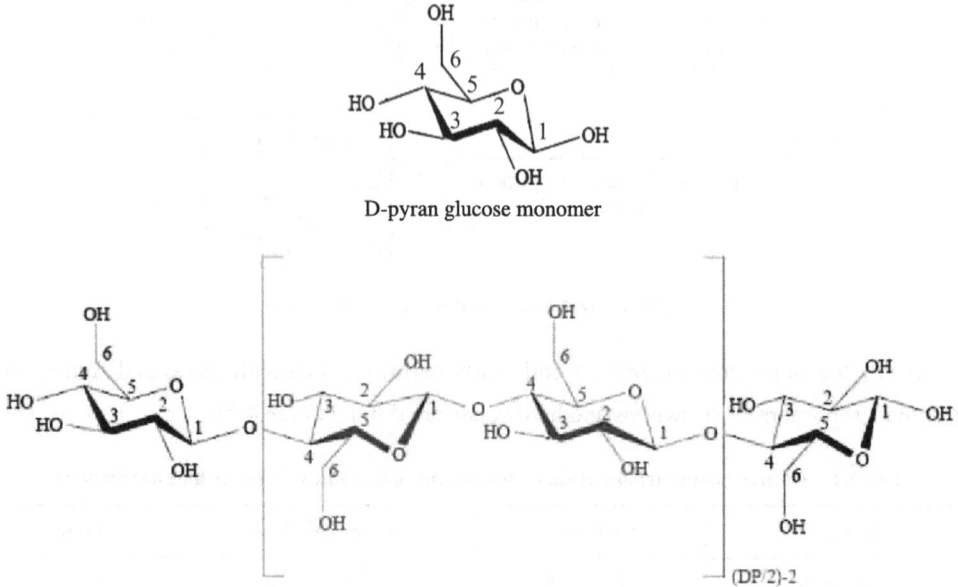

D-pyran glucose monomer

Fig.4.2   Molecular structure of cellulose

(2) Physical Structure of Cellulose

According to X-ray investigation, the aggregation of cellulose macromolecules is a two-phase structure (Fig.4.3). That is to say, cellulose is formed by alternating crystalline and amorphous regions. The molecules in the crystalline regions are regularly and closely arranged, mainly bonded via hydrogen bonds. However, the molecules in the amorphous regions are loosely arranged with poor regularity. There is no clear boundary between the crystalline and amorphous region, but only a gradual transition. Some of the cellulose molecules in the amorphous region form hydrogen bonds, while the remaining molecules are in the free state. Furthermore, cellulose moisture absorption occurs in the amorphous region.

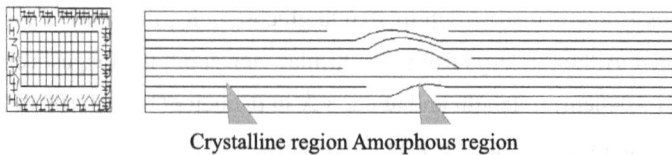

Crystalline region Amorphous region

Fig.4.3   Crystalline and amorphous regions within a microfibril

Crystallinity refers to the percentage of the crystallized area in the whole fiber. If the crystallinity is high, the tensile strength, bending strength, and dimensional stability of wood are high. In contrast, if the crystallinity is low, the above properties will inevitably decrease, and the hygroscopicity and chemical reactivity of wood will also increase.

(3) Main Properties of Cellulose

① Hygroscopicity of Cellulose　Some of the hydroxyl groups in the molecular chain of the amorphous regions of cellulose can form hydrogen bonds, while others are in a free state. Free-state hydroxyl groups are polar groups, which can easily absorb polar water molecules and form hydrogen bonds with them. This is the intrinsic reason why cellulose has hygroscopicity. The hygroscopicity depends on the series of the amorphous region: the hygroscopicity increases with an increase of the amorphous zone, in other words, the decrease of the crystallinity. If the hydroxyl groups of cellulose molecular chain were replaced, the hygroscopicity of cellulose would correspondingly change.

② Hydrolysis or Pyrolysis of Cellulose　When cellulose interacts with acid, the glycoside bond in macromolecules chain is unstable. With the appropriate hydrogen ion concentration and temperature, cellulose is hydrolyzed and converted into glucose in the end. Glucose is hexose and it can be further converted into alcohol through enzyme fermentation.

By diluting alkali at room temperature, the glycoside bond in cellulose gains high stability. When using concentrated alkali at room temperature, alkali cellulose can be formed.

When the temperature is below 140℃, the thermal stability of cellulose is good. However, cellulose will hydrolyze in water and oxidize in air. When the temperature is higher than 140℃, cellulose will turn yellow, and their solubility in alkali solution will increase. When the temperature is higher than 180℃, the degree of thermal decomposition will increase. If the temperature is higher than 250℃, cellulose will degrade violently.

4.1.2　Hemicellulose

Although glucose is the primary sugar produced in the process of photosynthesis, it is not the only one. Other six-carbon sugars, such as galactose and mannose, and five-carbon sugars, such as xylose and arabinose, are also manufactured in the leaves. These and other sugar derivatives such as glucuronic acid and glucose, are used within developing cells in synthesizing lower-molecular-weight polysaccharides called *hemicelluloses*. In contrast to the straight-chain polymer cellulose, most of the hemicelluloses are branched-chain polymers and generally made up of sugar units numbering only in the hundreds (that is, the degrees of polymerization are in the hundreds rather than thousands or tens of thousands).

Hemicellulose is a kind of copolymer composed of different sugar groups, which is an amorphous substance. Hemicellulose in terrestrial plants is composed of D-xylose, D-mannose, and D-glucose. Most hemicellulose contains short and numerous branched chains with low molecular weights.

Hemicellulose in hardwood is mainly composed of pentose, while that in softwood is mainly composed of hexose, with pentose being more stable than hexose. There are no chemical bonds between cellulose and hemicellulose, only secondary bonds.

Hemicellulose is the most hygroscopic, heat-resistant, and easy to decompose component among the three major components of wood.

4.1.3　Lignin

Lignin is a complex and high-molecular-weighted polymer built upon phenylpropane units. Although composed of carbon, hydrogen, and oxygen, lignin is not a carbohydrate nor even related to this class of compounds. Instead, it is essentially phenolic in nature. Many researchers have pointed out that there is a chemical connection between lignin and hemicellulose, and that hemicellulose molecules are mainly connected to lignin molecules through arabinose, xylose, and galactose, forming a lignin carbohydrate complex (LCC).

Lignin is a polymer with aromatic, non-crystalline, and three-dimensional structures. Its basic structural unit is phenylpropane. The three structural units of lignin are guaiacyl propane, syringyl propane, and p-hydroxy propane (Fig.4.4).

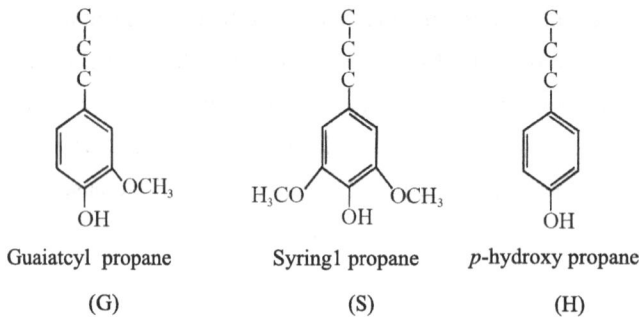

Fig.4.4　Three basic structural units of lignin

Lignin in softwoods is mainly composed of guaiacyl propane, while that in hardwoods is mainly composed of guaiacyl propane and syringyl propane. Lignin in herbaceous plants (such as bamboo) contains more p-hydroxyphenylpropane (Tab.4.3).

Lignin can also be used to identify softwood and hardwood trees using the color reaction of lignin (the mäule reaction). Firstly, wood samples are treated with a 1% potassium permanganate solution for 5 minutes. Secondly, it is washed with 3% hydrochloric acid and water. Finally, it is soaked in a concentrated ammonia solution. From here, yellow or yellow brown coloring indicates coniferous trees, while red or red purple coloring indicates broad-leaved trees. Additionally, lignin contains many chromogenic groups, such as benzene rings and vinyl groups, which affects the color of wood.

Tab.4.3 Different lignin contents in different materials

Materials	Lignin content (%)		
	G	S	H
Spruce	94	1	5
Pine	86	2	13
Beech	56	40	4
Bamboo	35	40	25

4.1.4 Extractives

Wood *extractive* is a general term for substances extracted from ethanol, benzene, ether, acetone or dichloromethane, and other organic solvents and water. There are three main types of wood extractives: aliphatic compounds, terpenes and terpenic alcohols, and aromatic compounds.

Wood extractives also affect wood color. The content of extractives is different for different tree species, resulting in different colored wood. In addition to this, the volatile extracts in wood also lead to different wood smells. Wood extractives affect the permeability, durability, adhesion, and finishing properties of wood.

4.2 Cell Wall Structure of Wood

Recall that a tree is sheathed by a thin cambial layer, which is composed of cells capable of repeated division. New cells produced to the inside of this sheath become new wood, and those moved to the outside become part of the bark. In this section, the chemical configuration of woody cell walls is examined, as are steps in the development of new cells.

4.2.1 Chemical Structure

The cell wall of wood is composed of cellulose, hemicellulose, and lignin. About 40 cellulose chains form an elementary fibril that is about 15~35 Å (1.5~3.5nm) wide (Fig.4.5). These elementary fibrils then aggregate into *microfibrillar bundles* with some assistance from hydrogen bonding in which hemicellulose plays a role. A proposed model of the arrangement of elementary fibrils (Fig.4.5) shows crystalline and amorphous regions and variations in spacing between elementary fibrils in hardwoods and softwoods.

Next, what of hemicellulose and lignin? The hemicellulose, probably somewhat selectively, interacts through hydrogen bonding with the cellulose and have been implicated in the aggregation of elementary fibrils into microfibrils. Hemicellulose, as noted earlier, is known to sheath the microfibrillar bundles. Moreover, hemicellulose is chemically linked to lignin macromolecules

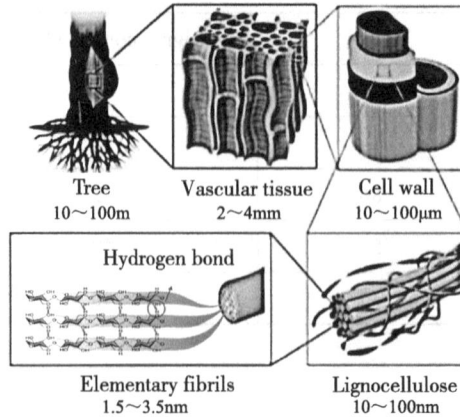

Fig.4.5   Multi scale structure of wood cell wall

and thus fulfill a particularly important function in maintaining cohesion between the architectural building materials of the wood cell wall. Although the precise nature of lignin continues to elude researchers, it is known that the aromatic rings tend to lie parallel to one another within the wood cell wall. However, neither lignin nor its chemical derivatives have ever been coaxed into a crystalline form from solution, whereas cellulose and hemicellulose crystallize quite readily.

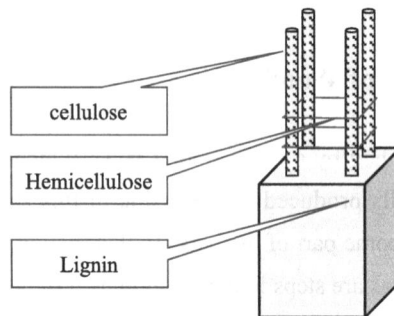

Fig.4.6   The existing state and function of the three components of the cell walls in wood

Figuratively speaking, cellulose acts as a skeleton material, which is equivalent to the steel bars among the reinforced-cement components (Fig.4.6). Hemicellulose infiltrates the skeleton material in the amorphous state and plays a role in matrix bonding. Hence, it is called a *matrix material*, which is equivalent to the fine wires tying the steel bars among the reinforced-cement components. Lignin is formed in the last stage of cell differentiation during the process of lignification. It penetrates the skeleton material and matrix material of the cell wall and hardens the cell wall. Therefore, it is called *crust material* or *hard solid material*, which is equivalent to the cement among reinforced-cement components.

4.2.2 Layering

Under a light microscope, the cell wall structure of wood can be divided into three layers (Fig.4.7): the intercellular layer (ML), the primary wall (P), and the secondary wall (S). Among them, the secondary wall can be divided into three layers: S_1, S_2, and S_3.

Intercellular layer ML: the middle layer between two cells, shared by two cells, very thin, mainly composed of lignin and pectin.

Primary wall P: the thickness is approximately 1% of the cell wall thickness.

Secondary wall: this wall layer, which is formed after cell maturation, can be divided into three layers. Among them, the thickness of the S_1 layer is 10%~22% of the cell wall thickness, and the microfibril angle (relative to the cell axis) is 50°~70°; the thickness of the S_2 layer is 70%~90% of the cell wall thickness, and the microfibril angle is 10°~30°; the thickness of the S_3 layer is 2%~8% of the cell wall thickness, and the microfibril angle is 60°~90°.

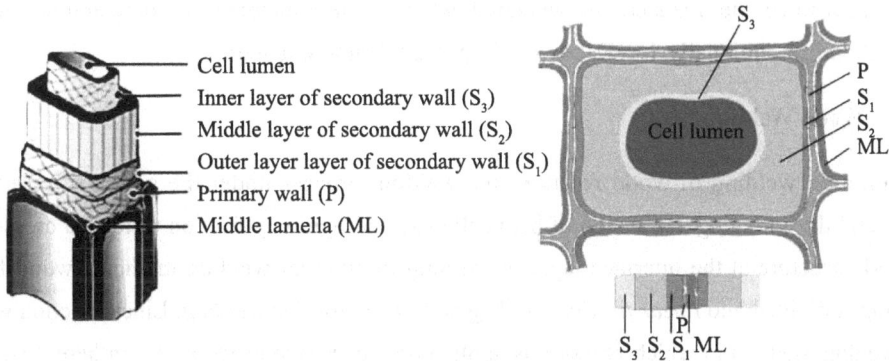

Cell lumen
Inner layer of secondary wall (S_3)
Middle layer of secondary wall (S_2)
Outer layer layer of secondary wall (S_1)
Primary wall (P)
Middle lamella (ML)

S_3
P
S_1
S_2
ML
Cell lumen

S_3 S_2 S_1 ML
P

Fig.4.7　Simulation structure of the wood cell wall layers and microfibrils

The pit is the channel between two cells, which is the sunken part of the secondary wall in the cell wall, including a single pit and a marginal pit (Fig.4.8). Pits are the channels through which water and nutrients are transported between adjacent cells. During wood seasoning and modification, the processing technologies of wood drying, antisepsis, fire prevention, and pulping are all related to the permeability of pits.

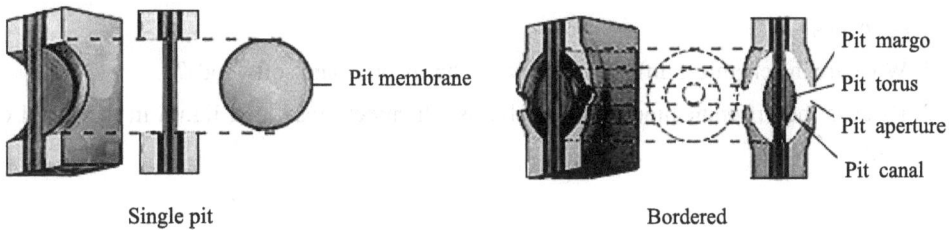

Pit membrane

Pit margo
Pit torus
Pit aperture
Pit canal

Single pit

Bordered

Fig.4.8　Pits

4.3　Examples of Wood Chemical Utilization

4.3.1　Pulp and Papermaking

Most paper is made from cellulose-rich wood. Lignin acts as a glue in plants, by tightly gluing fibers together. However, in papermaking, lignin is a harmful component. The main purpose of papermaking is to remove lignin as much as possible because any residual lignin will reduce the paper strength.

4.3.2　Manufacturing of Medium-density Fiberboard (MDF)

Papermaking requires pulping, and MDF preparation also requires hot milling. Yet, lignin does not need to be removed during fiber preparation. In the presence of high temperature and water, lignin can be softened and plasticized. When lignin is softened and, affected by a mechanical force, the fibers can be separated, which is the principle of pulping and hot grinding of fiberboard. Additionally, MDF is a very important furniture material.

4.3.3　Wood Welding

Friction welding of wood refers to the welding process under the action of an external force, and the softening and fusion of hemicellulose and lignin by friction to form a cross-linked network structure at the interface layer. According to different welding methods, wood friction welding is divided into *linear friction welding* and *rotary friction welding*. Linear friction welding is a welding method in which pressure is applied on the surface of wood. From here, two pieces of the wood base materials move under linear friction at a certain amplitude and frequency, while lignin and hemicellulose are blended on adjacent surfaces, thereby generating enough strength after the maintenance of pressure and cooling. Rotary friction welding refers to the welding method that inserts a rotary round wood tenon into a reserved hole under a certain pressure. This process forms a melting interface between the outside of the round wood tenon and the inside of the hole, generating a connection after further cooling.

❖ Exercises

1. What are the chemical differences between hardwoods and softwoods?
2. Discuss the different functions of cellulose, hemicellulose, and lignin in the wood cell wall.

Chapter ⑤ »»»
Physical Properties of Wood

The physical properties of wood are mainly determined by three characteristics: (i) the porosity or proportion of void volume, which can be estimated by measuring the density; (ii) the organization of the cell structure, which includes the microstructure of the cell walls and the variety and proportion of cell types (the organization of the cell structure is principally a function of species); and (iii) the moisture content (MC).

5.1 Wood Water and Density

5.1.1 Water in Wood

(1) The State of Water in Wood

In wood, its pores of different sizes form a capillary system. A capillary system that has a diameter larger than 0.2μm is called the *macro-capillary* system. It may include, for instance, cell lumen, intercellular space and pit cavity. The capillary system with a diameter of less than 0.2μm is called the *micro-capillary* system, for example, cell wall microfibril space. Water in wood can be categorized into *free water* and *bound water* based on its existing state (Fig.5.1).

Free water refers to the free-state moisture existing in the macro-capillary system of wood. The amount of free water is mainly determined by wood porosity, which affects wood quality, combustibility, permeability and durability, but has no effect on wood volume stability, mechanical, and electrical properties.

Bound water refers to the adsorption-state water existing in the micro-capillaries system of wood, that is, the water between the microfibrils of the cell wall. The amount of bound water has a great impact on the wood's physio-mechanical properties, wood processing, and utilization.

The water within the cell wall is held by adsorption forces, which are mainly hydrogen bonds. This is not to be confused with the absorption that takes place, for example, when a synthetic sponge soaks in water. In fact, absorption results from surface tension and capillary forces, leading to the bulk accumulation of water in porous wood. In contrast, adsorption involves the attraction of water molecules to hydrogen-bonding sites present in cellulose, hemicellulose,

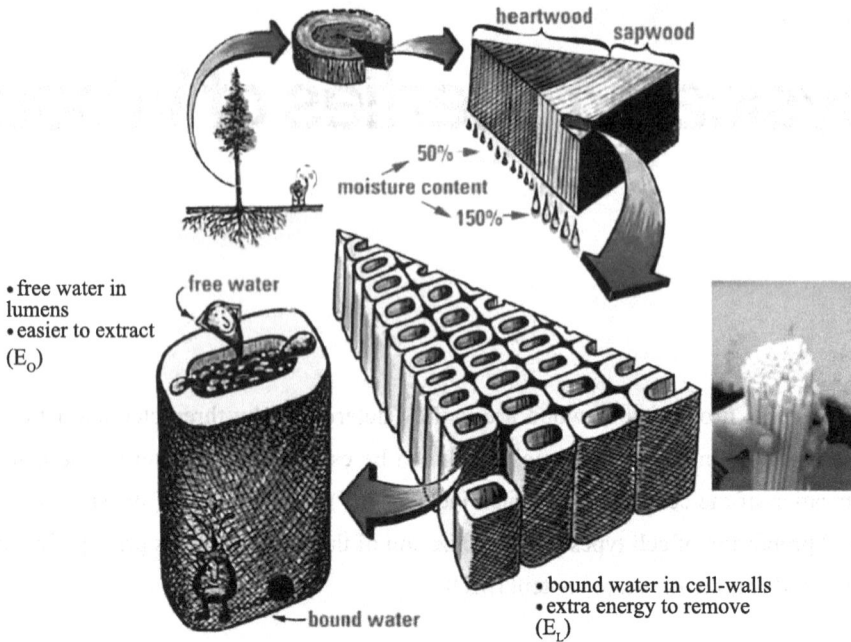

heartwood sapwood

50%
moisture content
150%

- free water in lumens
- easier to extract (E_O)

free water

- bound water in cell-walls
- extra energy to remove (E_L)

bound water

Fig.5.1　The states of water in wood

and lignin. Hydrogen bonding occurs on the hydrogen side of the hydroxyl (OH) groups, which can be found throughout the chemical components of wood.

(2) Calculation of Wood Moisture Content (MC)

Generally speaking, the terms *dry* and *wet* are used to qualitatively describe the different moisture of wood. Additionally, another term for wood *moisture content* (MC) can be introduced to quantitatively describe wood moisture. The MC is the weight of water contained in the wood, expressed as a ratio to the weight of wood.

Taking oven-dry (OD) wood mass, m_0, as the calculation basis, the moisture content, called **absolute MC**, is obtained as follows:

$$MC = \frac{m_1 - m_0}{m_0} \times 100\%$$

Where, m_1, m_0 represents the weight of the wood with moisture content (MC), and the weight of the OD wood respectively. Since the denominator is the wood's dry weight, but not the total weight, the MC calculated in this way can be over 100%. In some instances, such as in the pulp industry, the MC is based on the original weight, called **relative MC**:

$$MC = \frac{m_1 - m_0}{m_1} \times 100\%$$

Note that when calculating the MC, the amount of water is expressed as a percentage of the weight of the dry wood. This method of calculation is the accepted standard for all lumber, plywood, particleboard, and fiberboard products in the United States and in most countries.

(3) Moisture Content (MC) Measurement

According to the Chinese National Standard "*Method for Determination of the Moisture Content of Wood* (GB/T 1931—2009)", the dimension of wood samples should be 20mm×20mm×20mm. Each sample is immediately weighed and then placed in an oven heated to 101 to 105℃ [(103 ± 2) ℃] for 8 hours. In the process, two or three randomly selected samples are weighed to obtain the mass loss of wood. After that, the selected samples shall be weighed every 2 hours until the difference between the last two weighed masses of wood does not exceed 0.5% and the wood sample can be considered to be completely dry. Fig.5.2 shows balance and oven used to measure wood MC.

Balance   **Oven**

Fig.5.2　Equipment for measuring wood MC by gravimetric method

The major disadvantages of the oven-dry method are that: (i) it is a destructive test, requiring a sample to be cut from the piece; (ii) it can take several days to complete; and (iii) a few species contain volatile components other than water that can disappear during drying, thereby resulting in erroneously high moisture indications. However, for most situations, the oven-dry method is a more reliable indication of MC than that obtained by meters or other nondestructive methods. In fact, moisture meters and other nondestructive tests are calibrated using the oven-dry method.

(4) Equilibrium Moisture Content (EMC) of Wood

Wood is hygroscopic, and its MC is closely related to the relative humidity and temperature of the surrounding environment. When the water vapor pressure in the air is greater than that on the wood surface, the wood absorbs moisture from the air. This process is called *adsorption*. On the contrary, if the water vapor pressure of the air is less than that of the wood surface, the moisture in the wood evaporates to the air. This process is called *desorption* (Fig.5.3). Obviously, moisture absorption and desorption only refer to the adsorption and elimination of adsorbed water.

When wood is exposed to an atmospheric condition, its MC can finally achieve a relative constant value. That is to say, the moisture adsorption and desorption speed is the same. The wood MC at this state is so-called *equilibrium moisture content* (EMC). The EMC of wood mainly depends on the temperature and relative humidity (RH) of the atmosphere. Under certain temperatures and humidity conditions, wood EMC is constant. When the temperature is 20℃, the RH is 35%, while the EMC is 7%. When the RH is 70%, the EMC is 13%. Finally, when the RH is 65%, the EMC is 12%.

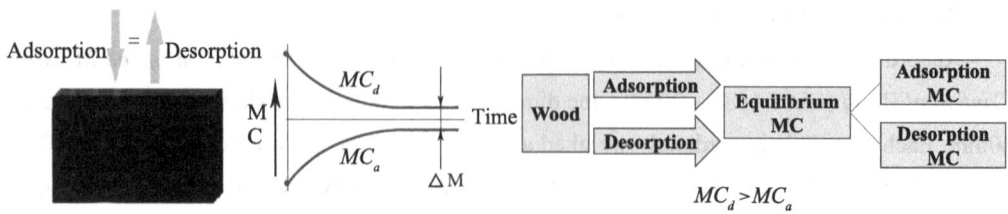

Fig.5.3 Adsorption, desorption and sorption hysteresis

Generally, at a given RH, the wood EMC can be attained through either adsorption or desorption. Achieving equilibrium from adsorption or desorption is represented in an exponential function of time (Fig.5.3). It may take weeks or even months to attain this equilibrium, due to the slow molecular rearrangements during the destabilization of wood cell walls.

The desorption EMC is measured by placing a wet wood sample under dry conditions. The absorption EMC is measured in the opposite direction (from the dry state to a relative wet condition). Under the same RH condition, desorption EMC is usually greater than absorption EMC ($\Delta MC = MC_d - MC_a$), which is called *sorption hysteresis*.

In practice, whether sorption hysteresis can be neglected or not depends on wood dimension and wood seasoning methods.

For fine wood, such as wood flour or thin wood like veneer, the sorption hysteresis is very small and can be ignored.

For air-dry wood, the sorption hysteresis is also small, and can roughly be assumed that the ΔMC is zero.

For kiln-dry wood, sorption hysteresis is very obvious, taking on a value of 2.5% in calculations. Thus, $EMC = MC_d = MC_a + 2.5\%$. During wood kiln drying, the final MC should be lower than $EMC - 2.5\%$.

MC has a great influence on wood properties. In the past, 15% was used as the standard MC in wood physical and mechanical property standards in China. In order to meet international standards, 12% is adopted as the standard MC today.

(5) Fiber Saturation Point (FSP)

Wood with different MCs can be divided into five categories, namely, green wood, wet wood, air-dry wood, kiln-dry wood, and oven-dry wood. The water states in the cell wall are shown in Fig.5.4.

Wood in a living tree generally has an MC of 30% or more, that is to say, the cell wall is fully saturated with water. The cell lumen generally contains some water, the amount of which varies greatly among trees and cells in the same tree. After the tree is felled, *green wood* can be obtained, and its MC may be above 50%. Sometimes, due to bacteria or physiological factors, wood is soaked in water to get an abnormally high MC. In turn, this wood is called *wet wood*. When green wood or wet wood is placed in the atmosphere, the moisture gradually evaporates. Finally, wood in equilibrium with the atmospheric humidity is called *air-dry wood*, and the MC at this time is called the EMC. Wood that is dried artificially can be called *kiln-dry wood*. If the drying continues to remove all water, the wood at this stage is called *oven-dry wood*.

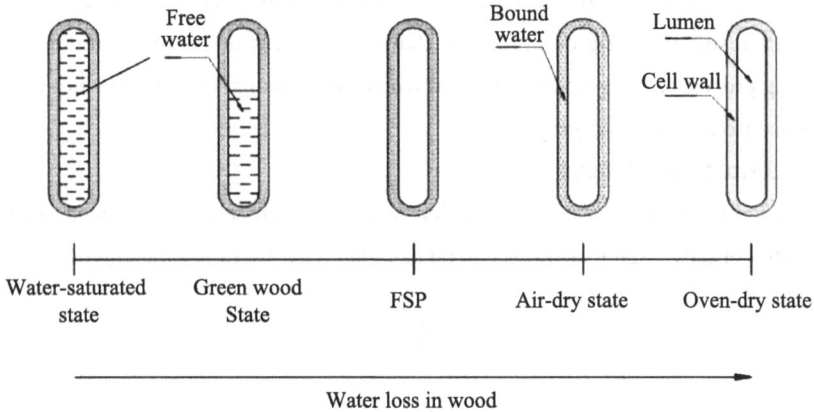

Fig.5.4 Different water states in wood

In Fig.5.4, the point at which all the free water in the lumen has been removed but the cell wall is still saturated is termed the *fiber saturation point* (*FSP*). Evidently, it is a critical state. FSP can range between 24%~42% and it is very difficult to measure. Thus, 30% is an average value and adopted as FSP in wood processing. Some believe that there is no such thing as FSP because the voids in wood go all the way to submicroscopic level. Therefore, there is no "full cell walls, empty lumen" case. FSP is the turning point of wood properties (Fig.5.5), so it is a very important characteristic of wood.

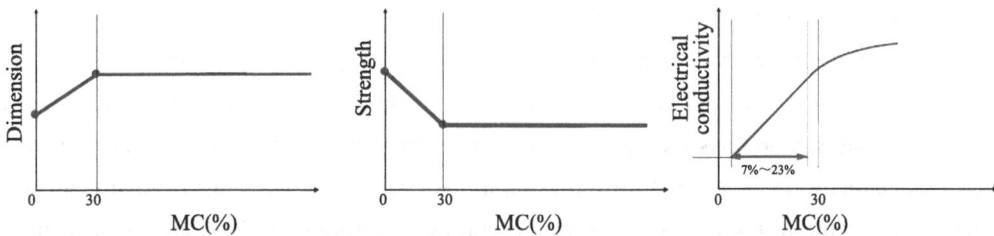

Fig.5.5 FSP and wood properties

5.1.2 Shrinking and Swelling of Wood

As mentioned above, at the living state or when MC is higher than FSP, wood is dimensionally stable. Below FSP, wood dimension changes as it gains or loses moisture, because the amount of bound water is related to the volume of the wood cell wall. When the amount of bound water increases or decreases, cell walls *swell or shrink*. Both swelling and shrinkage are considered under the phenomenon of "*hygro-expansion*". Wood is orthotropic in three primary directions, and therefore hygro-expansion is orthotropic as well.

As wood dries below the FSP, it loses bound water when it shrinks. Conversely, as water enters the cell wall structure, the wood swells. This dimensional change is a completely reversible process in small pieces of stress-free wood. In wood panel products, such as fiberboard and particleboard, however, the process is often not completely reversible. Partially, these characteristics result from the compression that wood fibers or particles undergo during the manufacturing process. In large pieces of solid wood, dimensional change may not be completely reversible because of internal drying stresses.

(1) Percentage Shrinkage

When wood is dried from wet to completely dry, the percentage dimension shrinkage with respect to radial or tangential directions can be calculated by the following equation:

$$\beta_{\max} = \frac{l_{\max} - l_0}{l_{\max}} \times 100\%$$

Where,

$\beta_{\max}$—Percentage dimension shrinkage of oven-dry wood with respect to radial or tangential direction;

$l_{\max}$—Radial or tangential dimension of wood with an MC above FSP, mm;

l_0—Radial or tangential dimension of oven-dry wood, mm.

And percentage volume shrinkage can be calculated by the following equation:

$$\beta_{v\max} = \frac{v_{\max} - v_0}{v_{\max}} \times 100\%$$

Where,

$\beta_{v\max}$—Percentage volume shrinkage of oven-dry wood;

$v_{\max}$—Volume of wood with an MC above FSP, mm^3;

v_0—Volume of oven-dry wood, mm^3.

Generally, from FSP to a completely dry condition, the percentage dimension shrinkage in the longitudinal, radial, and tangential direction are respectively 0.1%~0.37%, 3%~6%, and 6%~12%. Notably, wood has the most shrinkage (swelling) in the directions of annual growth rings (Tangentially, T). About half of the value in the direction across rings (Radially, R), and a very small value in the direction along the grain (Longitudinally, L). Percentage volume shrinkage is approximately equal to the sum of the three percent dimensional shrinkages.

(2) Coefficient of Shrinkage

The *coefficient of shrinkage* is used to determine the shrinkage allowance that should be reserved for wood processing. It is defined as the shrinkage variation per 1% MC change, which is the slope for the shrinkage-MC linear relationship, represented by K. Coefficients of shrinkage for tangential, radial, and longitudinal direction and volume are represented by K_T, K_R, K_L, and K_V.

$$K_{T、R、L、V} = (\beta_1 - \beta_2) / (MC_1 - MC_2)$$

Where,

β_1, β_2—Two different percentage shrinkages of wood below FSP, %;

MC_1, MC_2—Two different MCs below FSP, %.

The starting point of wood shrinkage is FSP, which is generally calculated as 30%. And the coefficient of shrinkage can be used to calculate the percentage shrinkage with any MC below FSP as follows:

$$\beta_W = K \ (30\% - MC)$$

In order to compare the difference degree of radial and tangential shrinkage in two different transverse directions, the *differential shrinkage* (*ratio of tangential shrinkage to radial shrinkage*) D is calculated by the following equation,

$$D = \beta_T / \beta_R = K_T / K_R$$

The average transverse shrinkage values for a number of species are shown in Tab.5.1.

Tab.5.1 Shrinkage values of wood from green to oven-dry MC

Species	Shrinkage (%)			Place of origin
	Radial	Tangential	Volumetric	
Chinese fir	1.23	2.91	4.20	China
Korean pine	1.22	3.21	4.59	China
Masson pine	1.50	2.96	4.66	China
Paulownia	1.47	2.69	4.53	China
Ash	1.97	3.53	5.77	China
White ash	4.9	7.8	13.3	USA
Quaking aspen	3.5	6.7	11.5	USA
Sugar maple	4.8	9.9	14.7	USA
North red oak	4.0	8.6	13.7	USA
Black walnut	5.5	7.8	12.8	USA
Western hemlock	4.2	7.8	12.4	USA
Loblolly pine	4.8	7.4	12.3	USA

(3) Bad Consequence and Control Method of Wood Shrinkage

As shown in Fig.5.6, the loss of moisture in wood leads to shrinkage. However, wood adsorbs moisture, which leads to swelling. In addition, when wood loses moisture, the internal stress is caused due to the uneven internal moisture content. If the internal stress is greater than the transverse strength, wood cracking will occur.

The control method of wood shrinkage are as follows:

① **Wood Surface Finishing**　Surface finishing can make wood water-proof or moisture-proof (Fig.5.7).

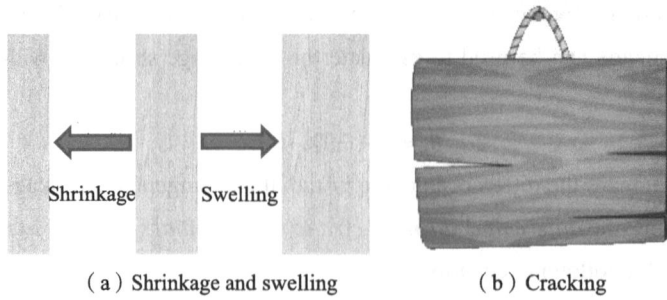

（a）Shrinkage and swelling   （b）Cracking

Fig.5.6   Bad consequence of wood shrinkage

Fig.5.7   Wood finishing

② **High-temperature Drying**   It is undertaken to make wood materials dry until its MC is below EMC and keep dimensional stability according to sorption hysteresis.

③ **Wood Acetylation**   Acetylation can effectively transform free hydroxyl groups into acetyl groups in wood, which can be achieved through chemical reaction of wood with acetic anhydride (Fig.5.8). The change from free hydroxyl groups to acetyl groups greatly reduces the water adsorption of wood. Hence, wood obtains its dimensional stability and super durability.

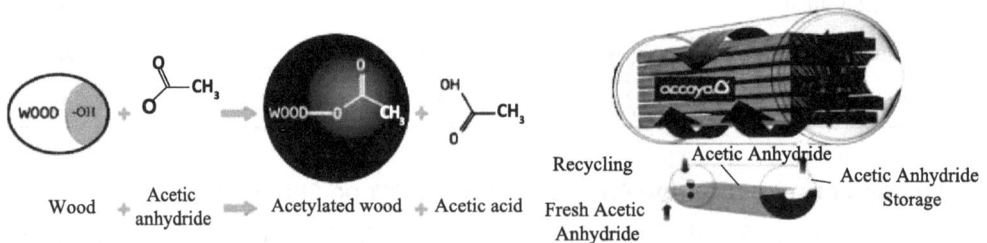

Fig.5.8   Wood acetylation (Diamond wood)

④ **Cross Lamination**   Wood veneer or dimension lumber can be cross-laminated and bonded together, that is, plywood or cross-laminated timber (CLT). Longitudinal shrinkage of wood is very small, and cross lamination of wood can suppress transverse shrinkage, and the hygro-expansion is reduced to a minimum. Additionally, wood strength along the grain direction is higher than that in the transverse direction, and cross lamination can make up for the disadvantages of low strength in the transverse direction of the wood (Fig.5.9).

Fig.5.9 Assembly of plywood and cross-laminated timber (CLT)

⑤ **Finger-joint Wood**   Wood, which are easy to deform and crack, are sawn into small square lumbers. Alternatively, small-diameter logs are processed into square lumbers. This entails gluing large square lumbers by means of seamless or seam finger joint. Wood deformation and cracking are practically eliminated (Fig.5.10).

Fig.5.10 Finger-joint wood

5.1.3 Density, Porosity of Wood

(1) Density

Density (D) can be described as the mass per unit volume, whose unit is g/cm^3 or kg/m^3. When discussing wood density, it should be noted that there is no universally accepted procedure for calculating the density of wood. For instance, although density is frequently expressed in terms of green weight and green volume when calculating weights for transportation or construction, this is not always the case. It is, therefore, important to ensure the basis of calculation when discussing wood density. It is a good practice to calculate density (the mass per unit volume) by determining the mass and volume at the same moisture content. The moisture content at which the density is determined should then be noted.

$$D = \text{mass/volume (at any given MC)}$$

(2) Cell Wall Density and Porosity

Because wood is porous, the measured volume of wood sample is its external volume, namely, the apparent volume (Fig.5.11). Thus, wood density can be called *volume weight* or *apparent density*.

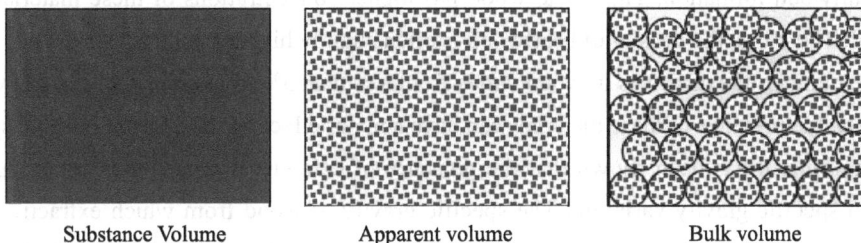

Substance Volume   Apparent volume   Bulk volume

Fig.5.11 Material volume

Because of its porosity, a piece of oven-dry wood is composed of the solid wood substance of its cell walls, the cell lumens which contains air, very small amounts of sap constituents (such as protein, minerals), and other substances (such as resins gums). An inverse relationship exists between the void volume of wood and density. This is because the density of void-free cell wall materials is taken from a low-density species (like basswood), tested for specific gravity, and compared with results of a similar test from a dense wood (such as hickory). Correspondingly, the two specific gravity values will be almost identical.

In general, it can be assumed that the density of dry wood cell wall material is 1.5 g/cm^3, meaning that specific gravity is 1.5. If a wood sample contained no cell lumina or other voids, as would be the case if it were completely crushed, it would have an OD specific gravity of 1.5. If it were 50% porous, its specific gravity would be 0.75. The approximate void volume of wood can be calculated by the following equation:

$$P(\%) = \left(1 - \frac{\rho_0}{\rho_{0w}}\right) \times 100\% = \left(1 - \frac{SG_{0d}}{SG_{0w}}\right) \times 100\%$$

Where,

P—Percentage void volume, %;

ρ_0—Bulk density of the wood based on dry volume, g/cm^3;

ρ_{0w}—Density of wood substance, g/cm^3;

SG_{0d}—Specific gravity of OD wood;

SG_{0w}—Specific gravity of OD wood substance.

(3) Calculation of Weight and Buoyancy

The weight of wood products can be easily estimated if the MC and specific gravity are known. The weight of the product should be calculated using the specific gravity that corresponds as closely as possible to the moisture content of the product at that time.

$$\text{Weight} = \text{volume} \times \text{specific gravity} \times \text{density of water} \times (1+\text{MC})$$

Wood often contains measurable quantities of extractives and infiltrates including terpenes, resins, polyphenols (such as tannins, sugars, and oils), as well as inorganic compounds (such as silicates, carbonates, and phosphates). These materials are located within the cell wall, where they are deposited during the maturation of the secondary wall and as remnants in the cell cavity after heartwood formation. Thus, heartwood has higher concentrations of these materials than sapwood. Therefore, the density of heartwood is often slightly higher than that of sapwood.

The number of extractives in wood varies from less than 3% to over 30% of the OD weight. Obviously, the presence of these materials can have a major effect on the density. In some species, including pine, it has been shown that the presence of extractives contributes significantly to observed specific gravity variation. The specific gravity of wood from which extractives have been removed tends to be more uniform than where the weight of extractives is included. In practical research, it is often desirable to determine the density of the wood without extractives. Both water and organic solvents are used for extraction when determining extractive-free density. Some of the bulking effect is lost when extractives are removed. So, in addition to weight loss, samples tend to show greater dimensional change with moisture fluctuations.

(4) Four Kinds of Wood Density

Since density is influenced to a large extent by the MC of the wood, the density of wood is related to the state of MC.

① **Basic density**　The basic density is calculated by dividing the oven-dry wood mass by the volume of wood saturated with water (green wood). *Obviously, basic density of wood is equal in numerical value to green specific gravity.*

$$\rho_b = \frac{m_0}{v_g}$$

Where,

ρ_b—Basic density, g/cm³;

m_0—Oven-dry wood mass, g;

v_g—Volume of green wood, cm³.

② **Density of green wood**　The density of green wood is obtained by dividing the mass of green wood by the volume of green wood.

$$\rho_g = \frac{m_g}{v_{max}}$$

Where,

ρ_g—Density of green wood, g/cm³;

m_g—Mass of green wood, g;

v_{max}—Volume of green wood, cm³.

③ **Density of air-dry wood**　The density of air-dry wood is obtained by dividing the mass of air-dry wood by the volume of air-dry wood.

$$\rho_a = \frac{m_a}{v_a}$$

Where,

ρ_a—Density of air-dry wood, g/cm³;

m_a—Mass of air-dry wood, g;

v_a—Volume of air-dry wood, cm³.

Due to different EMC in different areas, it is necessary to adjust the MC to a unified state in order to compare different species. Generally, the measured air-dry density of wood is converted into the value with an MC of 12%. The conversion formula is as follows:

$$\rho_{12} = \rho_a[1-(1-K_v)(MC-12\%)]$$

④ **Density of oven-dry wood**　The oven-dry wood density is determined by dividing the oven-dry wood mass by the oven-dry wood volume.

$$\rho_0 = \frac{m_0}{v_0}$$

Where,

ρ_0—Density of oven-dry wood, g/cm³;

m_0—Mass of oven-dry wood, g;

v_0—Volume of oven-dry wood, cm^3.

5.1.4　Measurement of Wood Density and Specific Gravity

(1) Direct Measurement

The wood sample should be processed into a standard cube with a size of 20mm × 20mm× 20mm, and the adjacent faces must be perpendicular to each other. Firstly, draw a mark at the center line of each relative surface of the sample. Secondly, measure the radial, tangential, and longitudinal dimensions with a screw micrometer, accurate to 0.01mm. Lastly, weigh it with a one thousandth scale to 0.001g.

The sample for air-dry density measurement is made of air-dry wood. After measuring the size, the air-dry wood mass is immediately weighed then put into the oven. The oven-dry wood mass of the sample is measured as mentioned previously. After the sample is dried, the oven-dry wood volume must be measured immediately. The air-dry wood density and oven-dry wood density can be calculated according to the relevant formula.

(2) Dewatering Measurement

In most cases, the dry weight is calculated by oven drying the sample, as would be done for MC calculation. Because high temperatures may drive off some of the extractives besides water, it is sometimes desirable to determine the MC by distillation. It is a process that involves condensing and weighing the collected vapor.

Sample volume may be obtained in a variety of ways. For a piece that is regular in shape, such as a section of lumber, the simplest method is to measure the dimensions as accurately as possible and calculate the volume. If the sample is irregular in shape, such as a tree cross-section or a wood chip, the volume can be obtained by the displacement method. The equipment used to do this is illustrated in Fig.5.12. The scale or balance records the weight of the fluid displaced. This value can then be converted to a volume by dividing the weight change by the density of the fluid used. The displacement procedure, using water as the fluid, works well with green materials because little water is adsorbed by the wood. When dry wood is immersed, the sample should be coated with paraffin wax or a similar substance so that water will not penetrate the block and produce an erroneously low volume determination. The use of a high-surface-tension, nonwetting fluid avoids the wetting problem with dry samples. However, since many such fluids (e.g. mercury) present safety hazards, the use of water and a waxy coating is usually more convenient.

(3) Fast Measurement

This method of determining volume involves the use of a graduated cylinder, as shown in Fig.5.13. In this case, the volume is simply the difference between the fluid level before and after immersion. This is a quick and simple technique, but it involves the same problem for dry sample as with the displacement method. The details of this method to determine the specific gravity of wood-based materials are discussed here. The sample is made into a cuboid of 2cm × 2cm × 20cm. The sample is required to be straight and regular, and the upper and lower end faces are parallel to each other. The whole length of the specimen is divided into 10 equal parts, which are marked as 0.1, 0.2, 0.3, …, 0.9. Then, one end of the specimen-which is marked 0.1 is immersed into the glass cylinder containing water, avoiding contact with the cylinder. At this time, the mark on the surface of sample with respect to the water surface, is the specific gravity (density) of the wood.

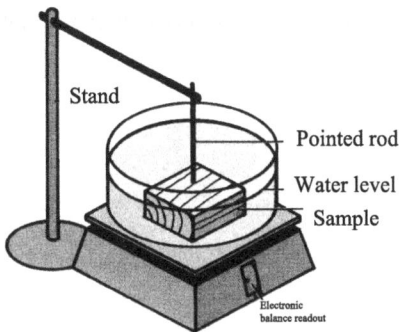

Fig.5.12　Determining volume by weighing before and after immersing the wood sample

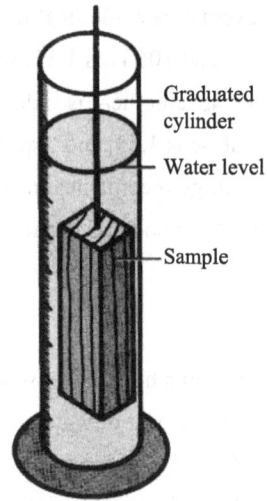

Fig.5.13　Determining volume by the difference in the water level

5.2　Other Physical Properties of Wood

5.2.1　Thermal Properties of Wood

(1) Specific Heat of Wood

The *specific heat* of a substance is the ratio of its thermal capacity to that of water at 15℃. If Q calories heat is necessary to raise the temperature of grams of a substance from T_1 to T_2 ℃ , the specific heat is defined as:

$$c = \frac{Q}{m(T_2 - T_1)}$$

The specific heat of wood is low, which is of importance for many technical processes, such as seasoning, impregnation, destructive distillation, and wood hydrolysis.

The true specific heat of oven-dry wood at the temperature is indicated by the following equation (Dunlap, 1912):

$$c_0 = 1.112 + 0.00485T$$

Where,

c_0—Specific heat of oven-dry wood, kJ/(kg · K);

T—Temperature, K.

In technical calculations for solid and liquid substances, mainly the average specific heat is important. Between 0℃ and 100℃, specific heat is expressed by the equation:

$$c = \frac{1}{100} \int_0^{100} \left(1.112 + 0.004\ 85T\right) \mathrm{d}T = 1.35 \mathrm{kJ/(kg · K)}$$

The average specific heat determined experimentally by Dunlap (1912) for twenty species between 0℃ and 106℃ is 1.35, with the minimum and maximum values being 1.325 and 1.409. The average specific heat is independent of both wood species and specific gravity. The specific heat of cellulose is 1.55, and that of charcoal is 0.669.

The average specific heat of wood varies considerably with MC. Assuming a simple addition effect of the dry wood and of the water, one can write:

$$c_{MC} = \frac{m_1 - m_0}{m_1} c_w + \frac{m_0}{m_1} c_0$$

Where,

c_{MC}—Specific heat of wet wood;

c_w—Specific heat of water;

m_1—Mass of wet wood;

m_0—Mass of oven-dry wood.

Since in wood and technology, the MC is based on oven-dry weight, namely, $MC = (m_1 - m_0) / m_0$. Further derivation leads to the following equation:

$$c_{MC} = \frac{(1 + MC)m_0 - m_0}{(1 + MC)m_0} c_w + \frac{m_0}{(1 + MC)m_0} c_0 = \frac{MC \cdot c_w + c_0}{1 + MC}$$

(2) Thermal Conductivity of Wood

Thermal conductivity is one of the most important thermal and humidity parameters of building materials, which is closely related to building energy consumption, indoor environment and many other thermal and humidity processes. The higher the thermal conductivity, the better the thermal transfer. Otherwise, the lower the thermal conductivity, the worse the thermal transfer. In engineering, materials with less than 0.23W/(m · K) are usually used as thermal insulating materials.

Wood and other cellulosic materials are poor conductors of heat. On the one hand, these are due to the paucity of free electrons, which are responsible for an easy transmission of energy (such as in metals). On the other hand, it is due to their porosity. Wood and wood-based materials are therefore used, to a large extent, as heat insulating materials in building construction, refrigerator cars, beer barrels, etc.

The thermal conductivity varies with the direction of heat flow with respect to the grain, density, type and quantity of extractives, defects, and especially moisture content. Under a steady-state temperature difference between faces of $(T_2 - T_1)$, thermal conductivity λ is the thermal energy Q per unit time t, which flows through a thickness Δx of a substance of a surface area A (Fig.5.14). Hence,

$$\lambda = \frac{Q \Delta x}{At(T_1 - T_2)}$$

Where,

Q—Thermal energy, J;

Δx—Thickness of wood, m;

A—Conducting heating surface area of wood, m^2;

$(T_2 - T_1)$—Temperature difference of wood surfaces, ℃ or K;

t—Conducting heat time, s.

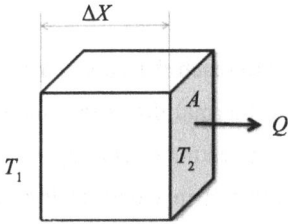

Fig.5.14 Heat conduction of wood

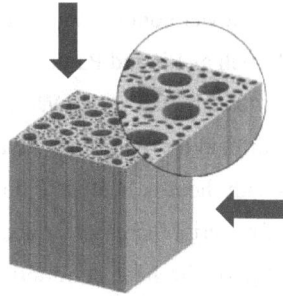

Fig.5.15 Porosity of wood and different heat flow directions

The thermal conductivity of a common structural wood is much smaller than that of metal. Tab.5.2 lists thermal conductivity of some common building materials.

Tab.5.2 Thermal conductivity of some common building materials

Materials	Thermal conductivity [W/(m · K)]	Materials	Thermal conductivity [W/(m · K)]
Structural softwood lumber (12% MC, transverse direction)	0.10~0.14	Concrete	0.9
		Mineral wool	0.036
Aluminum	216	Glass	1.0
Steel	45	Plaster	0.7

Owing to wood's ray tissue, the thermal conductivity of wood in the radial direction is about 5% to 10% greater than that in the tangential direction (Fig.5.15). In practice, both conductivities can be seen as the same asthe transverse conductivity of wood. Conductivity in the longitudinal direction is (in the MC range from about 6% to 15%) about 2.15 to 2.75 times the conductivity across the grain.

The conductivity of wood is nearly proportional to its density. The relationship between conductivity of oven-dry wood in the transverse direction and density can be described by the following equation:

$$\lambda = (5.18\rho_0 + 0.57P) \times 418 \times 10^{-4} = 0.217\rho_0 + 0.0238P$$

Where,

λ—Thermal conductivity, W/ (m · K) ;

ρ_0—Density of oven-dry wood, g/cm^3;

P—Porosity of wood.

The conductivity in the equation mentioned above is the superposition of conductivities of wood substance and dry air at room temperature. When $\rho_0 \to 0$ and $P \to 100\%$, $\lambda = 0.0238$W/(m·K). This number is the conductivity of dry air. And when $P \to 0$ and $\rho_0 \to 1.50$, $\lambda = 0.325$W/(m·K). This result is the conductivity of wood cell wall substance. So, the lower wood density, the higher porosity, and the lower conductivity and the better thermal insulation. This is the reason why low-density wood, such as Fir and Paulownia, is used to make pot cover.

(3) Thermal Diffusivity of Wood

Thermal diffusivity represents the ability of a material to reach the same temperature at each point the unsteady heat is heated or cooled down. A large thermal diffusivity indicates that the temperature of material increases or decreases rapidly during heating or cooling.

Knowledge of the heat conduction in wood is of interest in wood-drying and in wood-preserving. When analyzing the time-temperature relationship, it is firstly needed to calculate the diffusivity, or thermal conductivity constant.

The diffusivity factor α is as follows:

$$\alpha = \lambda / c\rho \ (\text{m}^2/\text{s})$$

Where,

λ—Thermal conductivity, W/ (m · K) ;

c—Specific heat, kJ/ (kg · K) ;

ρ—Density, kg/m^3.

Thermal performances of three common building materials is listecl in Tab.5.3.

Tab.5.3 Thermal performances of three common building materials

Materials	Specific heat c [kJ/ (kg · K)]	Thermal conductivity λ [W/ (m · K)]	Thermal diffusivity α (m^2/s)
Pine	2.51	0.16 (perpendicular to the grain)	0.11×10^{-6}
		0.35 (parallel to the grain)	0.24×10^{-6}
Steel	0.47	58	16.16×10^{-6}
concrete	0.88	1.60	0.71×10^{-6}

5.2.2   Electrical Properties of Wood

The electrical conductivity of wood in air-dry state is very small, especially when absolutely dry wood can be regarded as an insulator. Therefore, wood is one of the important insulating materials in transportation, electric power and other industries. However, if wood contains moisture, especially below the FSP, the higher the MC, the larger the electrical conductivity of the wood. The green wood is a conductor of electricity. Correspondingly, trees are easily knocked down by lightning in the rain (Fig.5.16).

Fig.5.16   Electrical conductivity of wet wood

▶ Exercises

1. White ash has an OD specific gravity of 0.61 and a green specific gravity of 0.55. What percentage of the total volume of dry wood is made up of cell wall substance? What will the weight of $1m^3$ of this wood be, if the moisture content is 38%? What will the moisture content of this wood be, if it is completely saturated with water (all voids filled)? At what moisture content will a log of white ash have zero buoyancy (i.e., just float)?

2. The average specific gravity of ponderosa pine at 12% MC is 0.40. What is the specific gravity of this ponderosa pine at the OD condition and when green?

Chapter ⑥ »»»
Mechanical Properties of Wood

As a kind of heterogeneous and anisotropic natural material, wood has many different mechanical properties from other homogeneous materials. The mechanical properties of wood include stress and strain relation, elasticity, viscoelasticity, strength, hardness, cleavage resistance, and wearing resistance, etc.

6.1 Basic Concepts of Wood Mechanics

6.1.1 Stress-Strain Relationship

Strain can be caused when a stress is applied to any solid body. In wood, if the stress does not exceed a level called the *proportional limit*, there is a linear relationship between the amount of stress and the resulting strain. When the stress is removed, the strain is completely recovered. The shape of a typical stress-strain curve for wood tested parallel to the grain is shown in Fig.6.1. Below the proportional limit, the ratio of stress to strain, that is, the slope of the linear relationship, is the so-called MOE. In compression and tensile tests, this ratio is sometimes termed *Young's modulus* to differentiate it from the MOE as determined by bending.

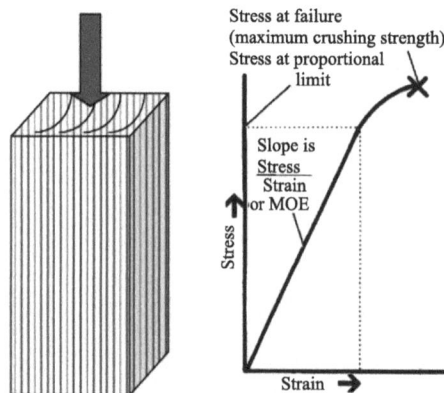

Fig.6.1　Relation between stress and strain in a typical compression parallel-to-grain test
(In general, the strain below the proportional limit is recoverable while strain above the proportional limit is permanent.)

In most of cases, above the proportional limit, the stress-strain curve cannot remain linear, but the strain can still recover. More specially, the strain is completely recovered when the stress is removed. We call this deformation in the elastic range *elastic deformation*. However, the difference between elastic limit and proportional limit is so small that they are usually not distinguished for wood. With an increase of external force, stress exceeds the elastic limit. After the external force is removed, the deformation of wood can only partially recover, and the residual partial deformation is called *plastic deformation or permanent deformation*.

Experimental results show that when shear stress does not exceed a certain limit, shear stress τ is directly proportional to its corresponding shear strain γ. Introducing the proportional constant G, we can obtain

$$\tau = G\gamma$$

This equation is called shear Hooke's law. And G is the shear modulus of elasticity, which is also an important mechanical index of material. This constant, has the same unit as shear stress.

6.1.2 Some Concepts of Wood Mechanical Properties

(1) Strength
Strength is the ability of a material to carry applied loads or forces without bending or breaking.

(2) Stiffness
Stiffness, or resistance to deformation, determines the amount a material is compressed, stretched, bent, or otherwise distorted by an applied load.

(3) Toughness
The toughness of a material reflects the ability of a material to absorb energy. Ash is an example of a tough wood that will not crack under a load. The opposite of toughness is brittleness. Glass is an example of a material that is brittle. Tough materials will resist impact forces well.

(4) Hardness
The hardness of a material is its ability to resist changing shape under force. Hard materials will resist wear and tear well, hence, they don't scratch or dent easily. Chisels, blades of planes, hammers and tool bits need to be hard, so they can keep their surface and shape for a long time. The blades of chisels are made of hardened steel so they can hold their edge well. If they were made of softer steel, they would go blunt rapidly.

Tabletops need to be hard so they will last and chopping boards need to be made from a hard material.

(5) Static load
The static load is the mechanical force applied slowly to an assembly or object.

(6) Impact load
When a moving object with a certain velocity strikes a stationary component, the velocity of the impactor changes greatly in a short time, that is, the impactor receives a great negative

acceleration. This shows that the impactor is subjected to a large force opposite to its motion direction. At the same time, the impactor also exerts a great force on the impacted component, which is called "impact force" or "impact load" in engineering.

(7) Vibration Load

A form of loading in which the magnitude and direction of forces are changed in turn.

(8) Long-term load

A form of loading in which the action of a force lasts quite a long time.

(9) Creep

Even if a load is small enough that there is no danger of ultimate failure, the member may continue to deflect or deform very gradually under constant stress. A common example is the gradual sagging of a shelf heavily loaded with books. This phenomenon is known as *creep*.

(10) Stress relaxation

Stress relaxation is the reduction in (relaxation of) stress when a material is loaded and then held at a constant level of strain.

6.2　Wood Elastic Constants

If a small rectangular sample is cut out from a trunk, at some distance from its pith, with a pair of faces tangential to the annual rings, this sample has three axes of symmetry. These are parallel to the longitudinal (L), radial (R), and tangential (T) directions, and are approximately perpendicular to one another (Fig.6.2). In reality

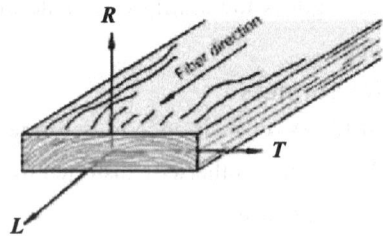

Fig.6.2　Principal axes and planes in wood

the radial axes in wood are not parallel but diverging, and LT-surface is not plane, but almost cylindrical. Nevertheless, it is customary, to assume three mutually perpendicular axes of elastic symmetry, which allow us to mathematically regard the system as a rhombic one.

A material with rhombic symmetry possesses 9 independent elastic constants and the compliance for this case states:

$$
S_{ij} = \begin{bmatrix}
\dfrac{1}{E_L} & -\dfrac{\mu_{LR}}{E_R} & -\dfrac{\mu_{LT}}{E_T} & 0 & 0 & 0 \\[2mm]
-\dfrac{\mu_{RL}}{E_L} & \dfrac{1}{E_R} & -\dfrac{\mu_{RT}}{E_T} & 0 & 0 & 0 \\[2mm]
-\dfrac{\mu_{TL}}{E_L} & -\dfrac{\mu_{TR}}{E_R} & \dfrac{1}{E_T} & 0 & 0 & 0 \\[2mm]
0 & 0 & 0 & \dfrac{1}{G_{RT}} & 0 & 0 \\[2mm]
0 & 0 & 0 & 0 & \dfrac{1}{G_{TL}} & 0 \\[2mm]
0 & 0 & 0 & 0 & 0 & \dfrac{1}{G_{LR}}
\end{bmatrix} \quad (i, j = L, R, T)
$$

Where, E_i is the MOE of wood, and the subscript i represents the directions of wood; μ_{ij} is the *Poisson's ratio* of wood, and the subscript ij represents the surfaces of wood, in which i represents the stress direction, and j represents the resulting strain in the other directions; G_{ij} represents the shear modulus of wood in three surfaces.

According to symmetry, the other three elastic constants of wood can be obtained by the following formula:

$$\frac{\mu_{ij}}{E_i} = \frac{\mu_{ji}}{E_j}(i, j = L, R, T)$$

Based on the elastic constants of wood listed in Tab.6.1, the following conclusions are obtained:

◎　Wood is a highly anisotropic material, and its MOE is different in three directions. The longitudinal MOE is more than ten or even dozens of times larger than transverse. The MOE in the radial direction is larger than that in the tangential direction.

◎　For shear modulus, it is the lowest in the cross-section of wood.

◎　MOE and shear modulus increase with an increase of wood density.

◎　The Poisson's ratio of wood is less than 1.

Tab.6.1　Independent elastic constants of typical softwoods and hardwoods

Materials	Density (g/cm³)	MC (%)	E_L (MPa)	E_R (MPa)	E_T (MPa)	G_{LT} (MPa)	G_{LR} (MPa)	G_{TR} (MPa)	μ_{RT}	μ_{LR}	μ_{LT}
Softwoods											
Spruce	0.390	12	11 583	896	496	690	758	39	0.43	0.37	0.47
Douglas Fir	0.590	9	16 400	1300	900	910	1180	79	0.63	0.43	0.37
Hardwoods											
Balsa	0.200	9	6274	296	103	200	310	33	0.66	0.23	0.49
Beech	0.750	11	13 700	2240	1140	1060	1610	460	0.75	0.45	0.51

6.3　Main Mechanical Properties of Wood

6.3.1　Tensile Strength of Wood

External force acts on wood, causing tensile deformation. The maximum stress of wood prior to tensile failure is called *tensile strength*. The tensile strength of wood can be divided into tensile strength along the grain and tensile strength perpendicular to the grain (Fig.6.3).

Tensile strength along the grain is the greatest among strengths of wood, which is two or three times the compression strength along the grain, and twelve to forty times the compression strength perpendicular to the grain. Shear strength along the grain is only 6% ~10% of tensile strength along the grain. Average tensile strength of wood along the grain is about 117.7~147.1MPa. Thus, wood is rarely broken by tension.

Tensile strength along the wood grain depends on the strength, length and orientation of wood fiber or tracheid. The length of fiber or tracheid is the main factor affecting wood strength. The longer the fiber and the smaller the filament angle, the greater the fiber strength. Although wood has a high tensile strength along the grain, it is seldom considered in architecture. Now why is that?

◎ Firstly, the variability and structural heterogeneity of wood have adverse effects on the tensile strength along the wood grain. For example, knots, cross grain, and other structural defects can greatly reduce tensile strength along the grain of wood.

◎ Secondly, shear strength along the grain is very low. As a building component, wood will be damaged due to shear or extrusion before high tensile strength along the grain is fully exerted.

◎ Thirdly, it is difficult to test tensile strength along the grain, and not easy to obtain reliable the data of tensile strength.

However, tensile strength perpendicular to the grain is much lower, which is only 1/40~1/30 of the tensile strength along the grain.

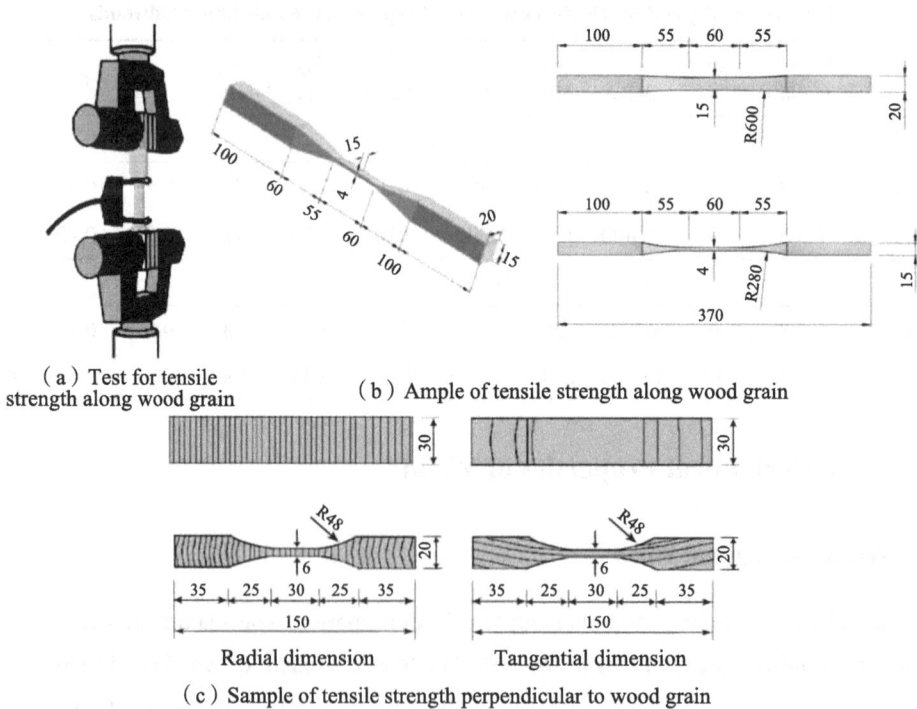

（a）Test for tensile strength along wood grain

（b）Ample of tensile strength along wood grain

Radial dimension   Tangential dimension

（c）Sample of tensile strength perpendicular to wood grain

Fig.6.3　Test for tensile strength of wood

The tensile strength along the grain of wood can be computed by dividing the ultimate load by the cross-section area of the slenderest position of the samples prior to testing.

$$\sigma_W = \frac{P_{max}}{bt}$$

Where,

σ_W —Tensile strength along the grain of wood with an MC, MPa;

P_{max} —Ultimate load, N;

b —Width of sample, mm;

t —Thickness of sample, mm.

The tensile strength along the grain of hardwoods with 12% MC can be computed by the following equation (The letter W represents MC):

$$\sigma_{12} = \sigma_W[1 + 0.015(W-12)]$$

If the wood is softwood and the MC ranges from 9% to 15%, $\sigma_{12} = \sigma_W$.

The transverse strength of wood with 12% MC can be calculated by:

Radial: $\sigma_{12} = \sigma_W[1 + 0.01(W-12)]$

Tangential: $\sigma_{12} = \sigma_W[1 + 0.025(W-12)]$

6.3.2   Compression Strength of Wood

According to the loading direction, the compression strength of wood can be divided into compression strength along the grain and compression strength perpendicular to the grain.

Referring to the Chinese National Standard *"Test Method of Compression Strength Perpendicular to the Grain"* (GB/T 1935—2009), the test method for the compression strength along the grain consists of applying the load along the grain, making wood fail at a uniform speed, to determine the compression strength (Fig.6.4). The compression-parallel-to-grain tests are conducted on 30mm × 20mm × 20mm samples.

$$\sigma_W = \frac{P_{max}}{bt}$$

Where,

σ_W —Compression strength along the grain of wood with an MC, MPa;

P_{max} —Ultimate load, N;

b —Width of the sample, mm;

t —Thickness of the sample, mm.

The compression strength of wood along the grain with 12%MC can be computed as follows:

$$\sigma_{12} = \sigma_W[1 + 0.05(W-12)]$$

In China, average compressive strength along the grain is about 45MPa. The ratio of the proportional limit and parallel-to-grain strength is about 0.7. This ratio is about 0.78 for softwoods, but 0.7 for low-density hardwoods and 0.66 for high-density hardwood.

Fig.6.4 Compression-parallel-to-grain test

In practice, the resistance of wood against perpendicular-to-grain compression is important for much building constructions and railway ties. Yet, one has to keep in mind that normally crushing strength across the grain does not exist. The wood only will be densified under the influence of compression force perpendicular to the grain. Thus, the proportional limit stress of perpendicular-to-grain compression is defined as the compression strength perpendicular to the grain. As shown in Fig.6.5, compression strength perpendicular to the grain has two loading methods, that is, total compression and local compression. According to the Chinese National Standard *"Method of Testing in Compression Perpendicular to Grain of Wood"* (GB/T 1935— 2009), the dimension of sample for total compression is 30mm×20mm×20mm, and it is 60mm× 20mm×20mm for local compression. The dimension of the centrally located rectangular steel plate for loading is 30mm × 20mm × 10mm.

(a) Total compression perpendicular to the grain   (b) Local compression perpendicular to the grain

Fig.6.5 Compression perpendicular to the grain.

With an MC, the compression strength of wood perpendicular to the grain can be calculated by:

$$\sigma_W = \frac{P}{ab}$$

Where,

σ_W —Proportional limit stress of wood perpendicular to the grain along the radial and tangential directions with an MC, MPa;

P —Proportional limit load, N;

a —Length of the sample or width of the steel plate, mm;

b —Width of the sample, mm.

To covert this strength of an MC to that of 12% MC:

$$\sigma_{y12} = \sigma_{yW}[1 + 0.045(W-12)]$$

6.3.3 Bending MOE and MOR of Wood (Center-point Bending)

When using a center-point-loaded test specimen with a rectangular cross-section (Fig.6.6), the flexure formula reduces to the following equation:

$$\sigma_W = \frac{3P_{max}l}{2bh^2}$$

Where,

P_{max}—The breaking (maximum) load, N;

l—The distance between supports (span), mm;

b—The width of the beam, mm;

h—The depth of the beam, mm.

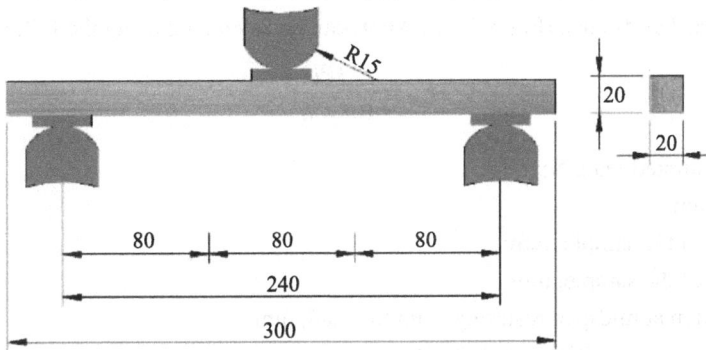

Fig.6.6   Bending MOE and MOE test (center–point bending)

For a test specimen loaded in bending with a concentrated load at the center of its span and supported at its ends, the MOE can be calculated using the following formula:

$$E_W = \frac{Pl^3}{4fbh^3}$$

Where,

P—Concentrated load, N;

l—Span, mm;

b—Width of the sample, mm;

h—Depth of the sample, mm;

f—Deflection at midspan resulting from the load, mm.

6.3.4 Bending MOE and MOR of Wood (Third-point Bending)

Bending modulus of elasticity (MOE) is defined as the ratio of stress and strain below proportional limit during wood beam bending. The maximum bending strength of solid wood and wood-based products is usually expressed in terms of the *modulus of rupture* (MOR). The MOR is calculated from the maximum load (load at failure) in a bending test, using the same static bending testing procedure for determining the MOE.

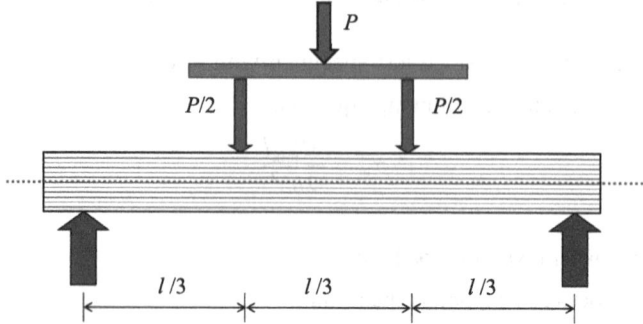

Fig.6.7　Bending MOE and MOE test (third-point bending)

For a test sample which is loaded in bending with a concentrated load at the third point of its span and supported at its ends (Fig.6.7), the MOE can be calculated using the following formula:

$$E_W = \frac{23Pl^3}{108 fbh^3}$$

Where,

P —Concentrated load, N;

l —Span, mm;

b —Width of the sample, mm;

h —Depth of the sample, mm;

f —Deflection at midspan resulting from the load, mm.

To convert it to that MOE with 12% MC:

$$E_{12} = E_W [1 + 0.015(W-12)]$$

To continue to load until its failure, the maximum load can be obtained and thus MOR can be computed.

$$\sigma_{bW} = \frac{P_{max}l}{bh^2}$$

Where,

P_{max} —Maximum load recorded during the test, N;

l —Span, mm;

b —Width of the sample, mm;

h —Depth of the sample, mm.

To convert it to that strength with 12% MC:

$$\sigma_{b12} = \sigma_{bW} [1 + 0.04(W-12)]$$

6.3.5　Shear Strength of Wood

According to the direction of external force acting on wood, the shear strength of wood can be divided into parallel-to-grain shear strength and perpendicular-to-grain shear strength

(Fig.6.8). In practical application, the phenomenon of perpendicular-to-grain shear is uncommon. Wood fiber is crushed and result in tensile effect, without pure shear in the transverse direction. Therefore, it is not used as a material property index.

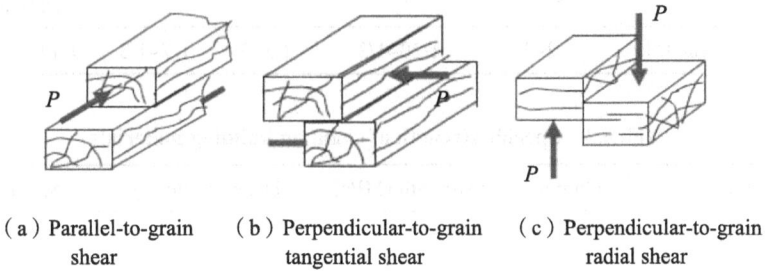

(a) Parallel-to-grain shear   (b) Perpendicular-to-grain tangential shear   (c) Perpendicular-to-grain radial shear

Fig.6.8 Shear test of wood

The shape and dimension of wood sample for parallel-to-grain shear is shown in Fig.6.9. Shear surface is tangential or radial surface and the length along the grain. The angle of the missing corner of the sample is 106°40′ with an allowance error of ±20′. The set-up is shown in Fig.6.10.

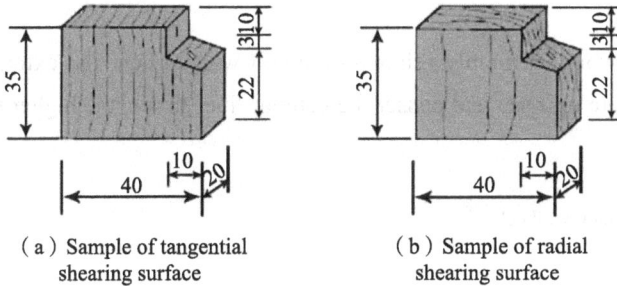

(a) Sample of tangential shearing surface   (b) Sample of radial shearing surface

Fig.6.9 Sample for parallel-to-grain shear

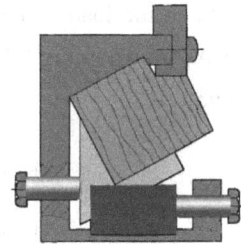

Fig.6.10 Set-up for parallel-to-grain shear

The shear strength is calculated by the following equation:

$$\tau_W = \frac{0.96 P_{max}}{bl}$$

Where,

τ_W —Shear strength, MPa;

P_{max} —Maximum load, N;

b —Width of the shearing surface, mm;

l —Length of the shearing surface, mm.

To convert it to that strength with 12% MC:

$$\tau_{12} = \tau_W[1+0.03(W-12)]$$

Relation between strengths of wood and specific strength of common building materials and listed in Tab.6.2 and Tab.6.3 respectively.

Tab.6.2 Relation between strengths of wood (Assuming that parallel-to-grain compressive strength is 1)

Compressive strength		Tensile strength		Bending strength	Shear strength		
Parallel-to-grain	Perpendicular-to-grain	Parallel-to-grain	Perpendicular-to-grain		Parallel-to-grain	Tangential surface shearing	Radial surface shearing
1	1/10~1/3	2~3	1/20~1/3	1.5~2	1/7~1/3	1/14~1/6	1/2~1

Tab.6.3 Specific strength of common building materials

Materials	Compressive strength (MPa)	Density (g/cm^3)	Specific strength
Low-carbon steel	400	7.8	51
Concrete	60	2.4	25
Pine (parallel-to-grain)	45	0.55	82

6.4 Influencing Factors of Wood Mechanical Properties

6.4.1 Wood Structure

When the load is applied to wood, it mainly relies on the cell wall to bear the external force. As shown in Fig.6.11, the more uniform and denser the cellular fiber tissue, the higher the strength.

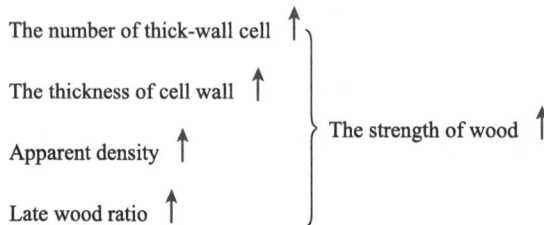

The number of thick-wall cell ↑
The thickness of cell wall ↑
Apparent density ↑
Late wood ratio ↑
} The strength of wood ↑

Fig.6.11 Wood structure affecting the mechanical properties

6.4.2 Moisture Content (MC)

The effect of wood MC on wood mechanical properties refers to the influence of wood MC below FSP on wood mechanical properties. When the MC is below the FSP, wood strength increases with a decrease of wood MC. This phenomenon is mainly due to an increase of the number of cellulose and lignin molecules per unit volume and an enhancement of the binding forces between molecules. When the MC is higher than the FSP, the free water amount increases, and the strength value no longer decreases and remains constant.

As shown in Fig.6.12, the change of wood MC has different influences on different strengths of wood. The bending strength and the compression strength along the grain are greatly affected, but the shear strength along the grain is less affected, whereas tensile strength along the grain is almost not affected.

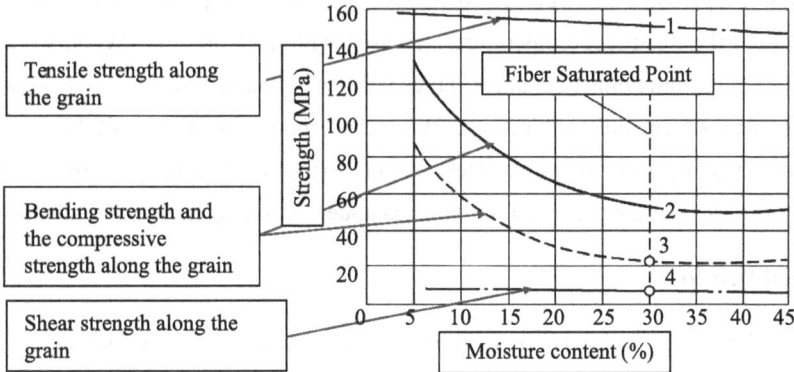

Fig.6.12   MC change and wood strength

According to the Chinese National Standard, strength of wood with 12% MC is defined as the standard strength. Thus, strength in other MC state should be converted based on the following equation:

$$\sigma_{12} = \sigma_w \left[1 + \alpha \left(W - 12\right)\right]$$

MC correction coefficient of wood strength is listed in Tab.6.4.

Tab.6.4   MC correction coefficient of wood strength

Strength	α value	Strength	α value
Parallel-to-grain tensile strength	0.015	Parallel-to-grain shear strength	0.03
Bending	0.04	Perpendicular-to-grain compressive strength	0.045
Young's modulus	0.015	Perpendicular-to-grain compressive MOE	0.055
Parallel-to-grain compressive strength	0.05	Hardness	0.03

Note that only hardwoods need MC adjustment for parallel-to-grain tensile strength.

6.4.3   Time-dependent

The resistance of wood to long-term load is different from that to short-term load. The constant velocity creep of wood under the action of an external force will produce a lot of rapid continuous deformation over a long time. The maximum stress that wood can bear under long-term load is called the *long-term strength of wood*. The long-term strength of wood is much smaller than its ultimate strength, which is generally 50% ~60% of the ultimate strength. All wood members are under the long-term action of load, so the effect of time on strength should be considered in timber design, and the long-term strength should be taken as a design basis.

6.4.4 Temperature

Wood strength decreases with an increase of ambient temperature. Results show that tensile strength and compression strength of softwoods decreased respectively by 10% ~15% and 20%~24% when the treatment temperature increased within 25~50 ℃ . When wood is kept at 60~100℃ for a long time, moisture and volatiles will evaporate and become dark brown, strength will decrease, and deformation will increase. When the temperature is over 140℃ , cellulose in the wood will pyrolyze, the color will turn black, and the strength will notably decrease. Therefore, it is not suitable to use wood structures for buildings under high temperatures for a long time.

6.4.5 Wood Defects

Wood defects, such as knots, cross grains, cracks, moths, decay, and so on, will cause the discontinuity and nonuniformity of wood structures, and thus affect the mechanical properties, and even bring adverse effects to the decoration.

⏵ Exercises

1. A 50mm×50mm×762mm clear, dry specimen of red oak is supported near each end of its 711mm long span and loaded at the center in a universal testing machine. A gradually increasing load is applied, and when the load reaches 6680N (below the proportional limit), the deflection at midspan under the load measures 6.6mm. What is the MOE of this specimen?

2. The same sample of red oak from exercise 1 is loaded flexurally to failure in a testing machine. The breaking load is found to be 9400N. What is the MOR?

Chapter ❼ »»»
Wood-based Materials

As a traditional material, wood has been used for a long time by human beings. Compared with other materials, wood has many unique properties, such as *porosity, anisotropy, swelling and shrinking, flammability*, and *biodegradability*. Therefore, main objectives from scientists and engineers includes how to improve uses of wood while avoiding its side effects to the greatest extent. With changes in natural resources and human needs, alongside scientific advancements, wood utilization has gradually developed from the traditional logs to timber, veneer, strand, fiber, and even nano-cellulose. These processes led to the formation of a huge family of new wood materials, such as *plywood, particleboard, fiberboard, laminated veneer lumber* (LVL), *glued laminated timber* (GLT), *scrimber, oriented strand board* (OSB) and decorative veneer, and other inorganic wood-based panels such as gypsum particleboard, cement particleboard, and *wood-plastic composite* (WPC), wood-metal composite, wood conductive material, and wood ceramic.

7.1 Conversion of Timber

7.1.1 Felling

Trees are cut down (felled) during the winter months when there is less growth in the wood. In Ireland, trees are often harvested by *clear felling* the cutting down of all trees in an area-as it is the most economical method. However, clear felling leaves large areas of poor, bare land and it affects the wildlife of the area. After felling, the trees are transported to sawmills to be cut into board of suitable size (Fig.7.1).

Fig.7.1　Felling trees, logging and sawing

7.1.2 Conversion

The process of cutting the logs into usable timber sizes is called *conversion*. Boards are cut and even the bark and small branches can be used as chip wood. The bark, which is removed from the logs, is turned into bark mulch.

By converting logs into boards, it allows:

◎ The wood to dry faster.

◎ The production of usable sizes and shaped wood.

◎ The quality of the timber to be seen and assessed.

Nowadays, the logs are cut using a large bandsaw. In the past, a saw pit would have been used. There are three methods for converting the logs:

◎ Through and through sawing.

◎ Quarter sawing.

◎ Tangential sawing.

(1) Through and Through Sawing

Through and through sawing is the fastest and most popular method of conversion. The logs are cut in parallel cuts in the direction of the grain. This form of conversion has certain advantages. It is also known as *plain sawing* or *slash sawing* (Fig.7.2).

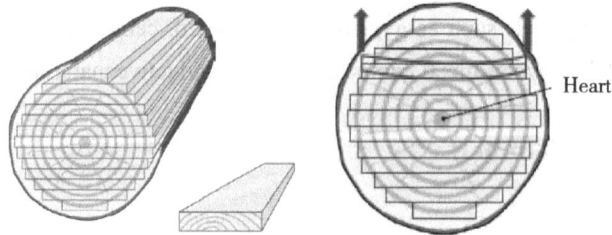

Fig.7.2　Through and through sawing

Advantages of through and through sawing: low cost; little waste; easy method as the board doesn't need to be turned; a fast method.

Disadvantages of through and through sawing: boards cut this way are likely to cup when drying; the boards show no particular grain pattern; less durable than other methods as there is a lot of sapwood in boards cut this way; it is not particularly suitable for structural timber.

(2) Quarter Sawing

In the quarter sawing method, logs are first quartered before cutting the boards, as shown below. This method of conversion displays as attractive grain figure when the ray cells are revealed (known as silver grain in oak). Cutting logs this way involves turning the log for each cut, so it is a labor-intensive work. Radial sawing involves cutting the quartered log in lines towards the center of the logs, as shown in the diagram. This process also reveals silver grain

(Fig.7.3).

Advantages of quarter sawing and radial sawing: attractive grain pattern is produced; the boards are more stable and shrink less; the boards wear more evenly (important in flooring, for example).

Disadvantages of quarter sawing and radial sawing: labor-intensive, as the log has to be quartered and then turned for each cut; an expensive method; more waste produced; the boards are not as wide as in plain sawing.

(3) Tangential Sawing

In the tangential sawing method, the cut is made at a tangent to the annual rings of the log. Timber converted in this way will highlight the flame figure that occurs in woods with distinct annual rings (Fig.7.4). Pitch pine and Douglas fir show the flame figure to great effect.

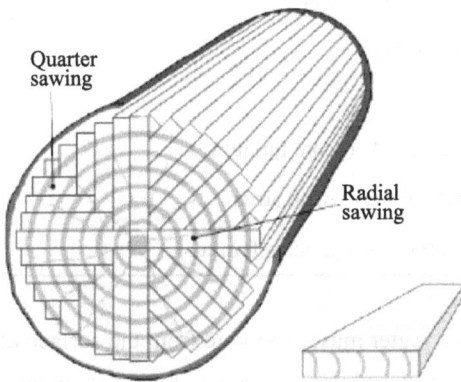

Quarter sawing

Radial sawing

Fig.7.3　Quarter sawing and radial sawing

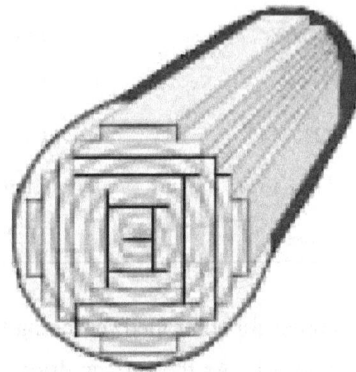

Fig.7.4　Tangential sawing

Advantages of tangential sawing: produces boards with flame figures; these boards season more quickly; the boards wear well; the boards can take a nail without splitting because of the position of the annual rings

Disadvantages of tangential sawing: boards converted this way tend to shrink and deform; the timber is prone to warping and cupping; it is expensive as the log is turned 90° for each cut.

7.2　Timber Seasoning

7.2.1　Water and Timber

The moisture content of a newly felled tree represents around 50% of the tree's total weight. When a tree is converted into log (green wood), this moisture starts to dry out quickly. The drying process must be controlled. Otherwise, defects will occur in the wood. Timber seasoning is the controlled drying of wood.

The aim of seasoning is to dry out the wood to a suitable moisture content of 20% or less.

(1) Reasons for Seasoning

◎ While wood will dry naturally, seasoning allows the process to be controlled.

◎ Seasoning helps to prevent the wood from splitting.

◎ Fungi do not attack dry timber.

◎ Dry wood is less likely to be affected by shrinkage or distortion.

◎ After seasoning, timber will be lighter, harder and stronger.

◎ Dry wood is easier to work with.

Suitable moisture content of wood in different applications is listed in Tab.7.1.

Tab.7.1　Suitable moisture content of wood in different applications

Moisture content	Situation
20%~22%	Limit of good air-seasoned wood
20%	Limit of dry rot occurring
16%	Outdoor furniture
12%~14%	Occasional heated areas, bedroom furniture
11%~13%	Living room furniture, well-heated areas
9%~11%	High degree of central heating, office furniture

In order for controlled seasoning to take place, water must evaporate from the surface of the wood gradually. As the surface dries, moisture from the center of the wood takes its place. It is important that the surface does not dry out too quickly or case hardening will result. The rate at which the wood dries depends on the relative humidity of the air around the wood.

(2) Relative Humidity

There is a certain amount of water vapor (moisture) in the air at all times. You cannot see this water vapor, but it affects, for example, the time it takes for clothes to dry on a line or for wood to season. There is significantly more vapor in the air during winter than in summertime. *Relative humidity* is the amount of moisture in the air at a given temperature, compared with the maximum amount of moisture the air could hold at that same temperature.

As wood is a substance that is affected by moisture changes (hygroscopic), it will release or absorb moisture to reach a balance (equilibrium) with the surrounding air. Wood, even in its finished form, for example a door, will soak up or let out water in order to reach this balance. This is called its *equilibrium moisture content* (EMC). Have you ever noticed how some outside doors will swell in winter and shrink in summer? This balance or equilibrium moisture content will change as the humidity of the air changes.

7.2.2　Seasoning

There are two methods of seasoning timber: natural or air seasoning and kiln drying.

(1) Natural (Air) Seasoning

Natural seasoning or air seasoning is where the planks or boards are laid down on large battens that are raised above the ground (Fig.7.5). The stack is made with timbers separated by stickers to give an air gap between the boards, which allows air to circulate up and down the stack.

Stickers or skids are small pieces of wood about 25mm in section, which help air to pass through the stack and, depending on how easily the timbers warp, can be spaced accordingly. The importances are as follows.

◎ The stack must be raised off the ground to keep it clean and dry.

◎ It must be covered with an overhanging lean-to roof so rainwater can run off.

◎ The stack must be on a dry, clean site.

◎ It must not be any more than two meters in thickness to allow air to flow through the stack. The free flow of air ensures that all the wood dries evenly.

◎ Timbers of the same species should be kept in the one stack.

◎ Boards of the same thickness and width should be kept together to keep the drying rate uniform.

◎ The ends of the boards should be protected from the sun, which may cause them to dry more quickly than the rest of the boards, causing the ends to split.

The air-seasoning method can only reduce the moisture content of a board to between 18% and 22% percent. Timber with this moisture level is suitable for joinery to be used out of doors. If a further reduction in moisture level is needed the boards must be kiln dried.

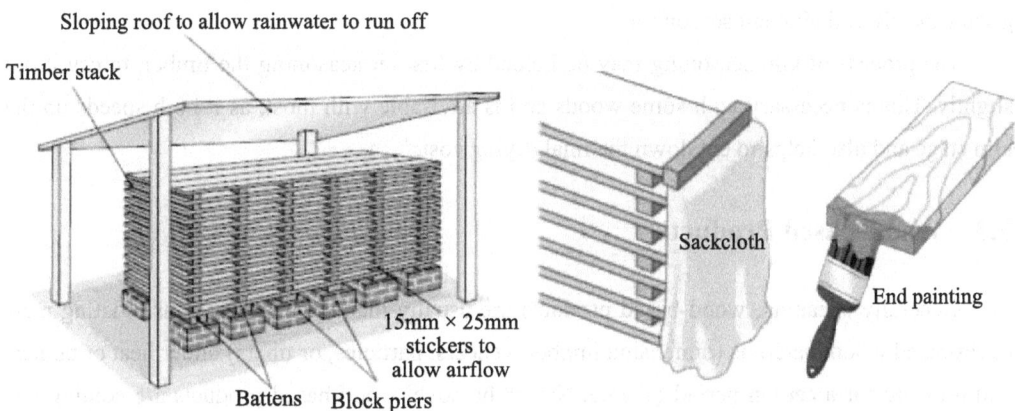

Sloping roof to allow rainwater to run off

Timber stack

Sackcloth

End painting

15mm × 25mm stickers to allow airflow

Battens　Block piers

Fig.7.5　Air seasoning and methods of preventing the ends of boards from splitting

(2) Kiln Drying

Another method for controlled wood seasoning is kiln (oven) drying (Fig.7.6). In these large kilns, the temperature, humidity, and drying rates can be controlled by the operator to achieve proper drying. The operator will have drying schedules (a guide for each kiln) for every types of timber and its particular thickness (as different thicknesses of wood dry at different rates).

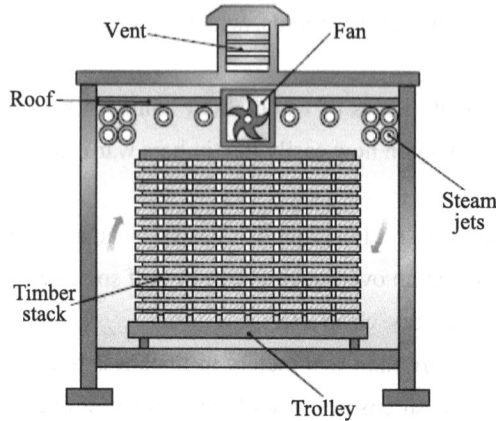

Fig.7.6 Compartment kiln

The timber is stacked on a trolley in a similar way to that used in the air-seasoning methods. The trolley runs on rails into the kiln. The steam jets on the walls, floor and ceiling pipe heated steam through the timber stack. The steam heats the wood but does not dry it. Once heated, the relative humidity of the kiln is reduced while the heat is maintained. This allows the moisture in the wood to evaporate gradually until the required moisture content is reached.

Fans circulate the air around the kiln and through the stack. Air vents allow moist wet air out of the kiln and fresh air in. Removing the moisture-laden air improves the efficiency of the kiln, giving speedy and efficient seasoning.

The process of kiln seasoning may be helped by first air seasoning the timber, to dry it out slightly. This is necessary with some woods and is advisable with most, as it both speeds up the kiln stage and also helps to cut down the final drying cost.

7.3　Wood-based Products

Generally speaking, wood-based products are usually manufactured by consolidating mats of resinated wood materials (dimension lumber, veneers, particles, or fibers) under heat or no heat and pressure for a certain period of time. Naturally, novel wood-based products are coming one after another based on the different combinations of wood elements. In turn, they can be classified into structural composites and nonstructural composites according to their end use. This textbook will concentrate on structural composites.

7.3.1 Structural Composites

(1) Lumber-based Engineered Wood Products (EWP)

Among engineered wood products, the group of structural wood products made from sawn timber can be defined as lumber-based EWP, including glue-laminated timber (GLT, which is also called glulam), cross-laminated timber (CLT), nail-laminated timer (NCL), and dowel-laminated timber (DLT) (Fig.7.7). Lumber-based EWPs can be prefabricated with precise dimensions and openings in a factory, which enables faster assembly with minimal construction waste. The key advantages of lumber-based EWPs include their light weight, high integrity, fast and easy on-site installation, excellent thermal and acoustic insulation, good durability, and quick investment return. With the advent of lumber-based EWPs, the market for mass timber construction has emerged throughout the world over the past several years. GLT and CLT are increasingly substituting solid timber as primary supporting structures.

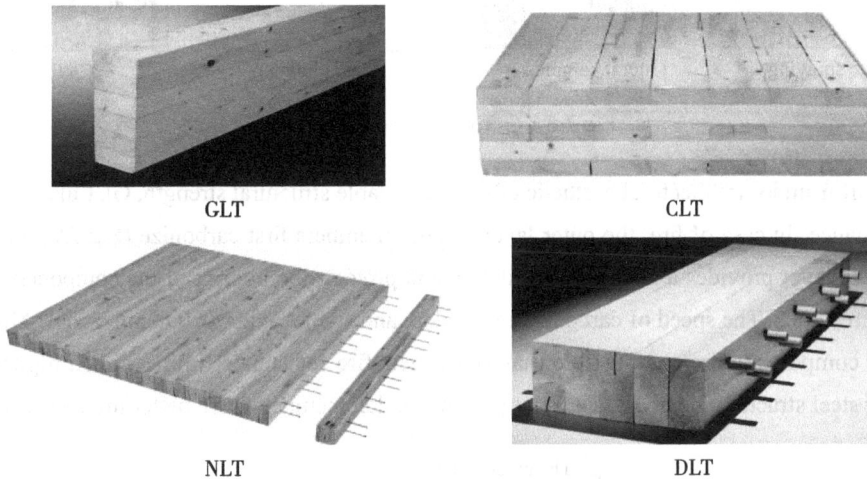

GLT   CLT

NLT   DLT

Fig.7.7 Lumber-based EWPs

① Glued-laminated Timber (GLT)

GLT, which has a long history in developed countries, is a traditional structural engineered wood product comprised of several laminates of dimension lumber that are bonded together with durable, moisture resistant structural adhesives. The grains of all laminations run parallel to the length of the member. GLT can be used in horizontal applications as a beam; in vertical applications as a column, or as curved, arched shapes. Furthermore, GLT is available in a range of strength classes.

After drying and grading, lumber layers of different grades are glued together in the longitudinal direction according to the different requirements for loading and uses. For a bending member, high-strength lumber is placed at the tension zone of a beam to improve its bearing capacity. The GLT production process has the following main steps (Fig.7.8).

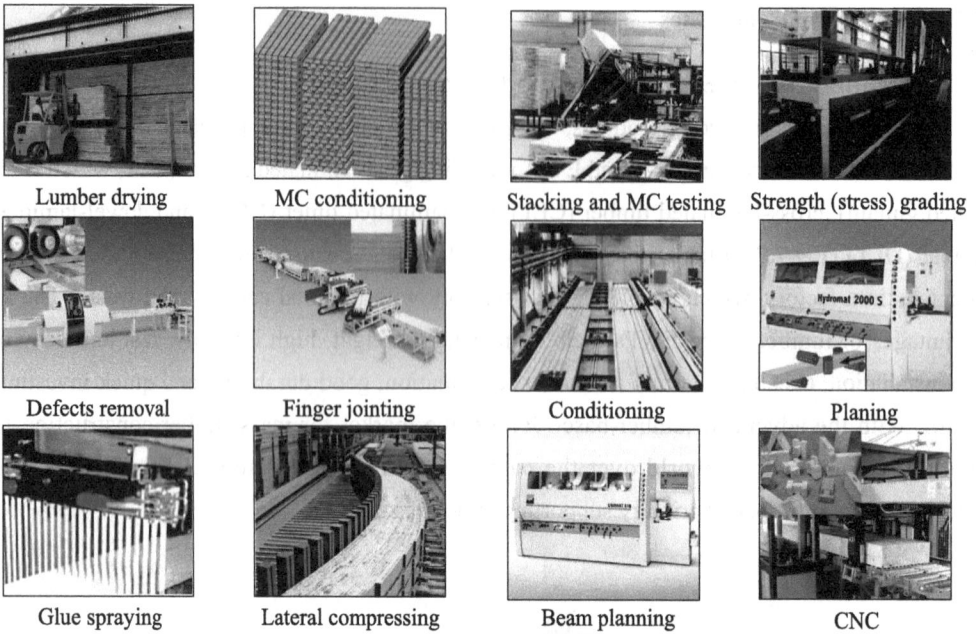

Lumber drying	MC conditioning	Stacking and MC testing	Strength (stress) grading
Defects removal	Finger jointing	Conditioning	Planing
Glue spraying	Lateral compressing	Beam planning	CNC

Fig.7.8 GLT production process

Apart from its architectural aesthetic effect and reliable structural strength, GLT also has good fire resistance. In case of fire, the outer layer of GLT members first carbonize (Fig.7.9), then this carbonized layer provides a good heat insulation and protects the interior of the components from further fire attack. The speed of carbonization is slow and stable, at about 0.635mm/min. Although wooden components are combustible materials, their fire resistance limit is much higher than ordinary steel structure members. Thus, large-size wooden members have better fire resistance.

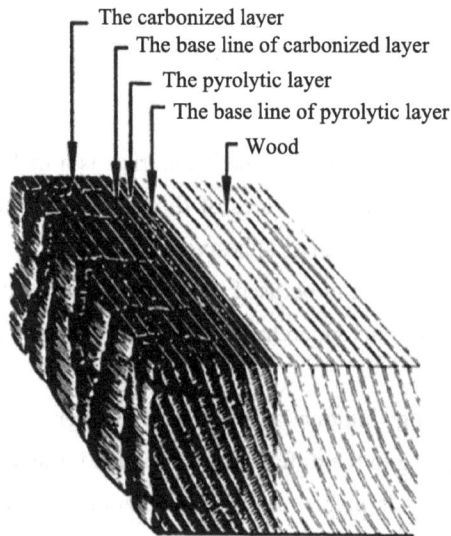

The carbonized layer
The base line of carbonized layer
The pyrolytic layer
The base line of pyrolytic layer
Wood

Fig.7.9 Cross-section of firewood

Furthermore, GLT can also be applied in some nonbearing fields, such as furniture-making and this kind of GLT is called nonstructural GLT (Fig.7.10).

Fig.7.10   Structural GLT and nonstructural GLT

GLT can be manufactured in different shapes, i.e., straight GLT, curved GLT, square-section GLT, rectangular-section GLT, and irregular-section GLT (Fig.7.11).

To make GLT beyond those commonly available for sawn timber in lengths, it must be end-jointed. In other cases, sawn timber needs to be cut to remove some defects and reassembled. End-jointing has four methods, that is, butt jointing, scarf jointing, tooth jointing, and finger jointing.

End-jointing is the process of connecting two pieces of lumber end-to-end (Fig.7.12). Because the end grain of the wood is composed of mostly open pores and little surface area, very little tensile strength can be developed by simply butt jointing the ends of wood. In turn, it is not used in the manufacturing of bending and tensile components. As a means of overcoming this hurdle, scarf jointing was developed. To scarf joint wood, a low-angle cut is machined into the ends of lumber, with mated pieces that are glued together under pressure. The result is a longer lumber than what is otherwise achieved with tensile strengths set at 90% of that of solid lumber. In comparison to scarf jointing, tooth jointing is suitable for fast manufacturing. Yet, it cannot reach a strength as good as scarf jointing. To minimize waste associated with end-jointing, finger jointing has been developed. In this case, the low-angle cut is folded back upon itself numerous times, giving the appearance of multiple fingers. Currently finger jointed lumber is used for structural members.

As shown in Fig.7.13, GLT can be made into symmetrical and asymmetric components according to their requirements. No.1, No.2, No.3 and T.L are different laminate grades. For bending members, the most important zone of the cross section is the edge tension zone. In asymmetric GLT beams, the wood grade at the edge of the tension zone is higher than that in the compression zone, which can effectively improve the utilization of wood. Therefore, the bending strength of the asymmetric glued beam is different between the tension zone and the compression zone.

Round-section GLT

Arc GLT

Square-section GLT

Irregualr-section GLT

Fig.7.11　Different shapes of GLT

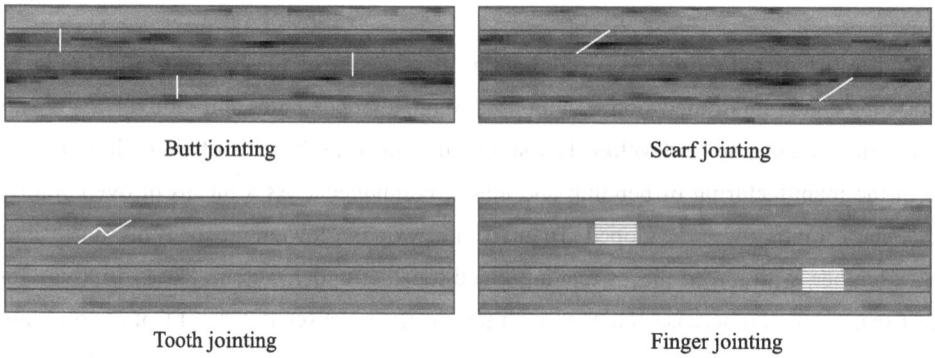

Butt jointing

Scarf jointing

Tooth jointing

Finger jointing

Fig.7.12　End jointing of lumber

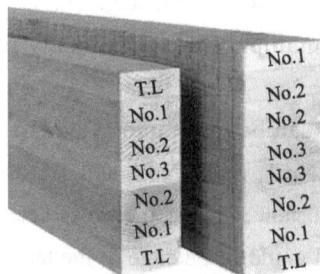

Fig.7.13　symmetrical and asymmetric GLT

　　GLT does not change the structural characteristics of wood, but its uniformity of material property is better than solid wood. Therefore, GLT can be used as a solid wood alternative in all structural or nonstructural fields (Fig.7.14).

Fig.7.14　Application of GLT

② Cross-laminated timber (CLT)

The CLT consists of at least three layers of structural lumber boards or SCL stacked crosswise at an angle (typically at 90°) and glued or stapled together (Fig.7.15). The CLT has the advantages of a high prefabrication rate, convenient transportation and fast installation, and low damage to the site environment. Thus, it is considered the best substitute for traditional building materials.

The crosswise layup method of CLT makes it possible to fully use the material characteristics of high tensile strength of wood in the direction of the grain and high compression strength in the transverse direction. To meet the special needs of CLT application, double-layer lumber boards can be placed in a parallel-to-grain direction for superior mechanical properties in the direction of the grain. Alternatively, Canadian scientists proposed a box-based CLT system to achieve a more diverse and complex structural purpose [Fig.7.15(c)]. This system effectively reduced the self-weight of CLT panels by ensuring a certain bearing capacity and became more cost effective.

The concept of CLT was developed in Austria in the 1970s and 1980s. The first modern CLT mill was established in Europe in the late 1980s. In 1993, the first CLT building was erected in Switzerland. In the past 10 years, Austria, Germany, and Italy have carried out lots of research on

CLT. While this product is well-established in Europe, Canada, the United States, New Zealand, and Japan, lots of related research work have been carried out, greatly promoting the applications of CLT. CLT can be widely used as wall, roof and floor panels, and also as a main body structure or decks for a bridge.

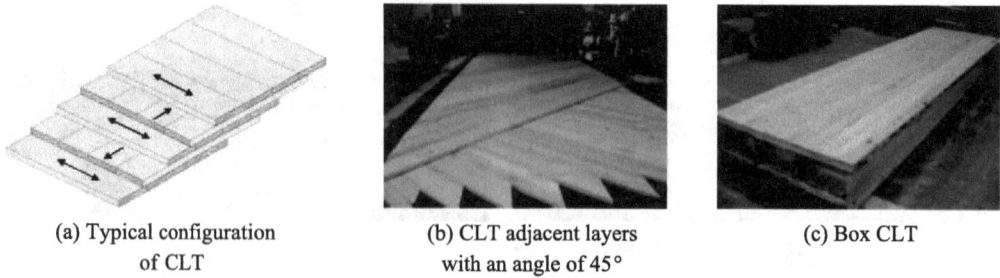

(a) Typical configuration of CLT

(b) CLT adjacent layers with an angle of 45°

(c) Box CLT

Fig.7.15　Different structures of CLT panels

Fig.7.16 shows a schematic representation of a typical CLT manufacturing process, which involves the following nine basic steps:

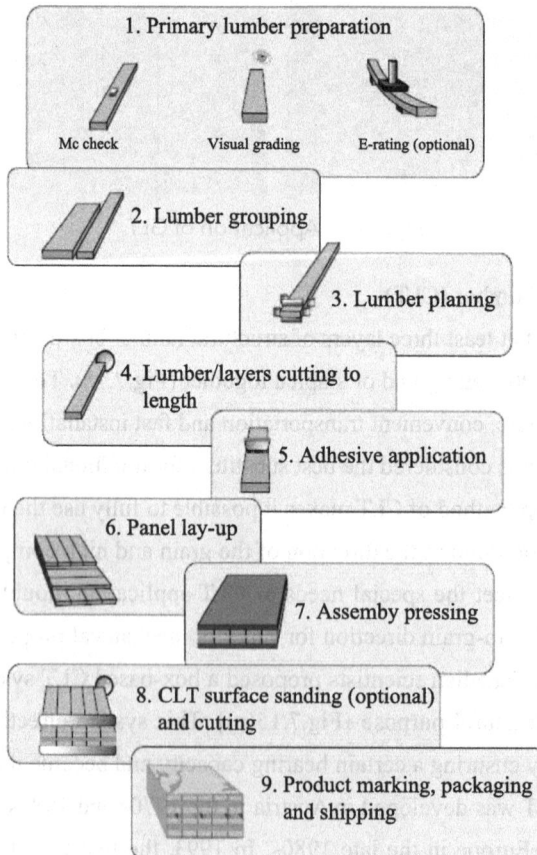

Fig.7.16　The manufacturing process of CLT

The key to a successful CLT manufacturing process is consistency in the lumber quality and control of the parameters that affect the bonding quality of the adhesive.

Lumber stock may be selected based on the grade of the CLT panel to be produced. That is, for appearance grade CLT, the outermost layer(s) may have specific visual characteristics for aesthetic purposes. Some manufacturers may produce a so-called composite CLT, by face bonding wood composites or engineered wood products, such as oriented strand board (OSB), plywood and LVL, to the CLT (Fig.7.17).

An 18-story student residence with a frame-core tube system, Brock Commons, was located at the campus of the University of British Columbia. The core tube was cast-in-place using reinforced concrete. The first floor of the building was a concrete frame structure. For the 2nd to the 18th floors, CLT panels were adopted as the horizontal force-bearing members and the GLT columns as the vertical force-bearing members, altogether connected by steel members.

The World Society for High Rise Buildings and Urban Human Settlements (CTBUH) announced that the Norwegian Mjøstårnet, which was completed in March 2019, is the tallest wooden structure building in the world (Fig.7.18). The building is 85.4m in height and has 18 floors. In order to avoid the problem of excessive horizontal displacement under wind load, a precast concrete floor slab is used in the upper part of the building.

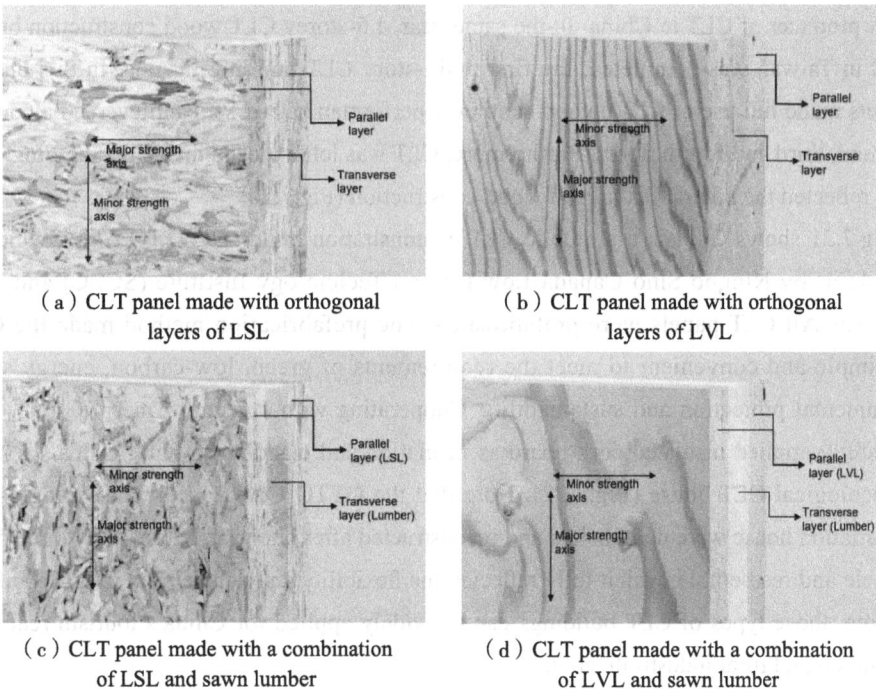

(a) CLT panel made with orthogonal
layers of LSL

(b) CLT panel made with orthogonal
layers of LVL

(c) CLT panel made with a combination
of LSL and sawn lumber

(d) CLT panel made with a combination
of LVL and sawn lumber

Fig.7.17　Composite CLT

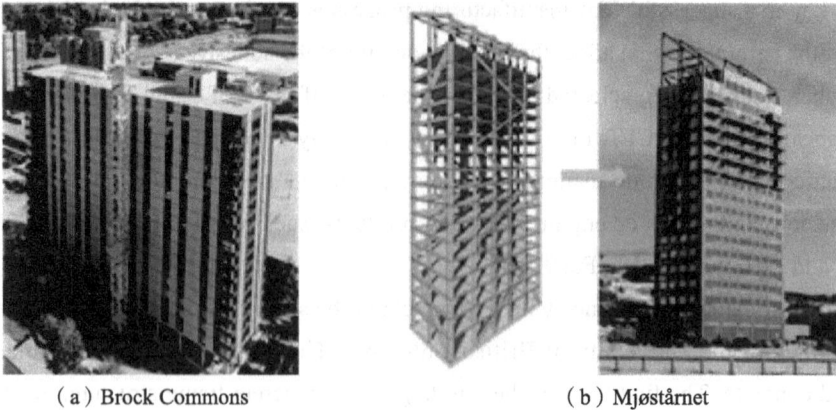

(a) Brock Commons　　　　　　　　　(b) Mjøstårnet

Fig.7.18　High-rise CLT buildings

Owing to the incompleteness of China's specifications and the public's lack of awareness towards multi- and high-rise wood buildings, CLTs were predominantly used in low-rise buildings in China. Additionally, these buildings were mainly built for demonstration purposes.

In March 2014, a two-story wood construction building with a hybrid light-frame and CLT structure in China was reported for demonstration in Qian'an City, Hebei province (Fig.7.19). Its CLT materials were manufactured by Qian'an City Big Tree Industry Co., Ltd., which was the first producer of CLT in China. In the same year, a 5-storey CLT wood construction building located in Taiwan was completed, the first multi-story CLT building in Asia. In this building, architects made full use of CLT's good cantilever performance, and set cantilevered balconies on the second, third and fourth floors. Furthermore, CLT was left exposed inside the building, which largely reflected the natural features of wood construction (Fig.7.20).

Fig.7.21 shows China's first public CLT demonstration building (OTTO Café), which was jointly built by Ningbo Sino-Canada Low Carbon Technology Institute (SCLC) and Tongji University. All CLT panels were prefabricated. The prefabrication method made the OTTO Café simple and convenient to meet the requirements of green, low-carbon, energy saving, environmental protection and sustainability. Cooperating with Portugal Amerin Group, SCLC successfully applied insulated cork board as insulation and cladding to build China's first two-story ecological CLT house (Fig.7.22). Note that the OTTO Café and the all-ecological CLT demonstration house were demolished and reconstructed after one year. It took only three days to dissemble and reassemble, which fully reflected the flexibility and adaptability of CLT buildings. Therefore, those types of CLT buildings can be widely applied for China's tourism real estate, urbanization and rural industrialization.

Moreover, Zhongyi Scientech Timber Structure Co., Ltd, one manufacturer of CLT located in Shandong Province, also carried out some engineered applications of CLT. The business building of the Binzhou Administrative Center was one case, which was designed to use composite CLT elements combined with wooden beams and columns.

Fig.7.19 CLT construction building in Hebei

Fig.7.20 CLT construction building in Taiwan

Fig.7.21 OTTO Café

Fig.7.22 Ecological CLT demonstration house in Zhejiang

③ Comparison between CLT and GLT

CLT is the latest addition to the EWP family which offers unlimited size and potential for larger, taller, more energy efficient and durable building systems. The species used for GLT and CLT manufacturing are generally the same: mainly softwoods, including Norway spruce (*Picea abies*), white fir (*Abies alba*), Scots pine (*Pinus sylvestris*), European larch (*Larix decidua*), Douglas fir (*Pseudotsuga menziesii*), and western larch (*Larix occidentalis*). Spruce-pine-fir (SPF) and Norway spruce are the most common species for GLT and CLT manufacturing, respectively, in North America and Europe. Other species including hardwoods can also be used for CLT. This large variation in wood selection is possible because of the solid composition and homogenization of panel properties within CLT.

According to current manufacturing methods, CLT panels must be formed in a large-scale press under pressure in the vertical and horizontal directions. When polyurethane (PUR) is used, a 1~2h period of cold press is necessary. Obviously, the manufacturing efficiency for CLT panels is higher than that for GLT panels. From an application standpoint, CLT is primarily used for floors, walls and roofs. It is processed less often for beam and column members. By contrast, GLT is mostly used for beam and column applications.

(2) Veneer-based Engineered Wood Products (EWP)

① Structural Plywood and Laminated Veneer Lumber

In most of types of plywood, the grain of every other layer is applied parallel to the first. The grain orientation of adjacent veneers lies at right angles. However, with aminated veneer lumber (LVL), all of the veneer sheets are oriented parallel to the long axis of the product (Fig.7.23). Douglas fir, larch pine, southern yellow pine, yellow poplar, aspen, western hemlock, lodgepole pine and spruce are main species groups used for the manufacture of structural plywood and LVL in North America. The Adhesive commonly used in structural plywood and LVL is phenolic resin.

To begin production, all logs are debarked. For plywood and LVL, logs are cut to length on a slasher saw into bolts, ~2.64 m. Most softwood veneer blocks are heated prior to peeling. Heating softens the wood and knots, thereby reducing cutting power consumption and producing a higher volume of smoother, higher-quality veneer. Steaming, soaking in hot, slightly alkaline water, spraying with hot water, or a combination of these methods are all used in various situations to obtain the increased wood temperature required. The objective is to heat logs to a suitable temperature so that veneer will be cut. Most softwoods are heated to a core temperature of 50~60°C. The exact heating time required depends upon the density of the wood, the diameter of the block, the bath temperature, the initial wood temperature, and the temperature required for a satisfactory cut. Virtually all veneer in structural veneer is peeled on lathes. Rotary lathes are designed to produce continuous ribbons of green (undried) veneer that is subsequently clipped to recover usable veneer widths. A veneer slicer is used for manufacturing a lower-production, higher-quality veneer generally utilized in nonstructural panels. Most structural veneers are peeled to a target thickness of 2.5~4.2mm.

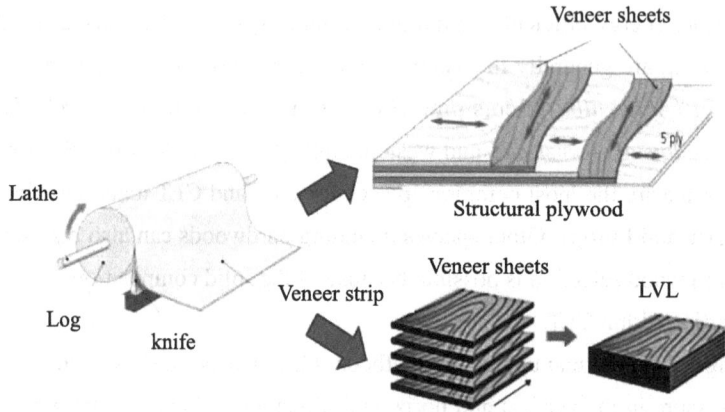

Fig.7.23 Veneer-based EWP

Most structural veneers are first graded ultrasonically. The ones with the highest grades (stiffest, strongest, and densest) are routed to LVL production. The ones with a lower grade are then visually graded for use in plywood.

② Parallel Strand Lumber (PSL)

Parallel strand lumber (PSL) is manufactured by gluing long veneer strands together with the grain of each strand oriented to the length of the final product (Fig.7.24). While this design loses the transverse dimensional stability associated with OSB, it maximizes the product's bending strength. Veneer used in PSL is sourced from structural plywood or LVL mills. The first or initial veneer removed from logs as they are rounded-up on a lathe (as the logs' taper is removed) does not come off in full-length pieces. Rather it comes off in short and irregular pieces, which are difficult to incorporate into plywood or LVL. However, this veneer is generally of excellent wood quality, has no heartwood, making it readily gluable and treatable with preservatives and having only few defects, such as knots. This veneer is processed through a clipper and a trimmer to make the long strands necessary for PSL. Strands are ~20mm wide, 4mm thick, and up to 1 m long. Strands are then dried to a moisture content of 3%~5%. PF-based adhesive is then applied to the long strands. Billets are pressed with a high-frequency (radio or microwave) energy to cure the phenolic resin adhesive in a continuous-type caterpillar press that slightly increases the density of the wood.

Fig.7.24   PSL product and application

(3) Strand-based Engineered Wood Product (EWP)

Oriented strand Board, OSB, is a relatively new manufactured board (Fig.7.25). It is a material that looks similar to chipboard but has many of characteristics of plywood. Wooden strands of flakes are processed from the tree log and are then bonded together under heat and pressure, using a synthetic adhesive and wax. The strands are aligned in two outer layers and an inner core positioned at right angles to the outer layers, creating a three-ply effect. This gives the board its transverse strength.

The board comes in a standard size and is suitable for a wide variety of uses. It is used in packing cases, flooring, furniture manufacture and timber-framed buildings. It has moderate resistance to moisture and fungal attack.

Laminated strand lumber (LSL) is similar to PSL, except that it is made with thinner and wider strands and it uses different resin binder. Aspen, yellow poplar, basswood, or other low-density hardwood flakes are produced on a modified disk flaker. Strand-production equipment is similar to that used in the manufacture of OSB. However, the strands are about twice as long, which is around 30cm. Green strands are dried and then screened. Broken or short strands are continually removed

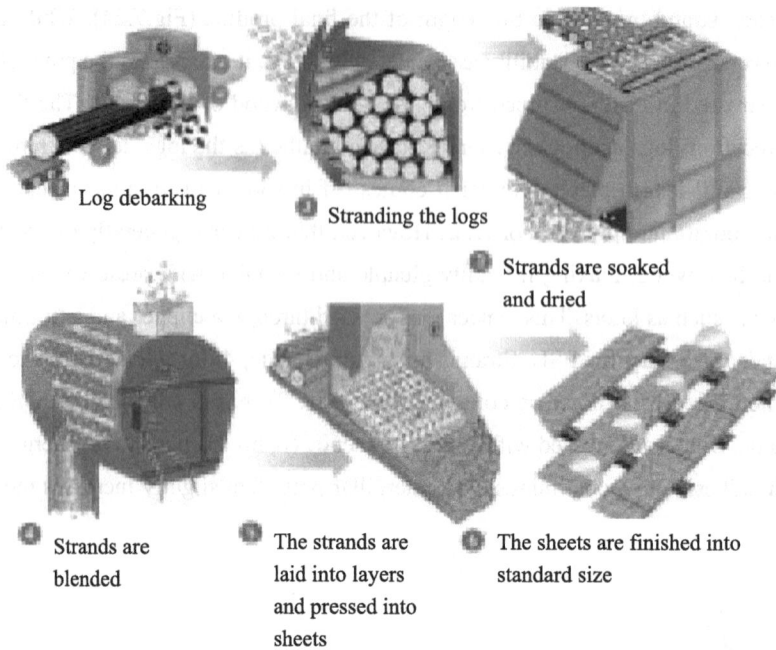

Log debarking

Stranding the logs

Strands are soaked and dried

Strands are blended

The strands are laid into layers and pressed into sheets

The sheets are finished into standard size

Fig.7.25　The main steps of making OSB

from the process. Then the strands are blended with resin and formed, like OSB, into a loose mat. This product is manufactured with pMDI (polymeric diphenylmethane diisocyanate) resin adhesive to provide a waterproof, fast-curing, light-colored product. The use of pMDI reduces press cycle time in comparison to phenolic adhesives. The billets, 25~150mm thick, are cured in steam-injection presses. Steam injection further reduces press cycle time and minimizes density gradients through the billets' thickness. Because diphenylmethane diisocyanate (MDI) utilizes moisture in its curing reaction, strands can contain relatively high levels of moisture, on the order of 15%, enhancing dryer throughput 32mm thick panels can be cured in 60 seconds.

Oriented Strand Lumber, OSL, is produced from oriented flakes, in much the same way as the faces and core of OSB. In this case, however, all strands are oriented in one direction. OSL differs from LSL in that its flakes are shorter, about 15cm long, the same length as those used in OSB. Specialized curved press platens allow for numerous architectural shapes to be produced (Fig.7.26). These products have a ready market in architectural windows and doors, furniture parts, and other specialty applications.

7.3.2　Nonstructural Composites

(1) Blockboard

Blockboard is a strip core board. It is made up of thick strips or battens of solid wood, usually softwood, glued together and covered with veneers on both sides just like plywood (Fig.7.27). It is a thicker board, usually 12~25mm wide, as thin boards would be difficult to

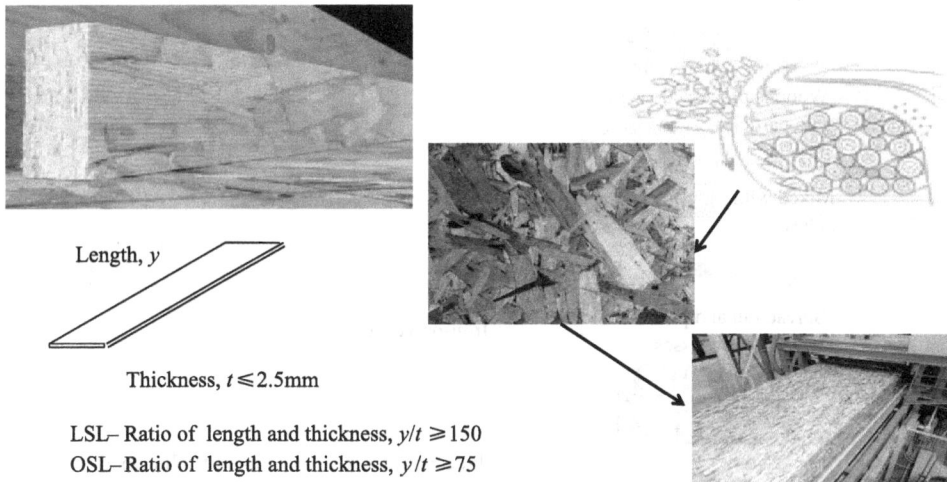

Length, *y*

Thickness, *t* ≤ 2.5mm

LSL– Ratio of length and thickness, $y/t \geqslant 150$

OSL–Ratio of length and thickness, $y/t \geqslant 75$

Fig.7.26   LSL and OSL

manufacture. The grain of the facing veneer runs 90° to the direction of the solid strips. The strips in the center are cut from lower quality timber.

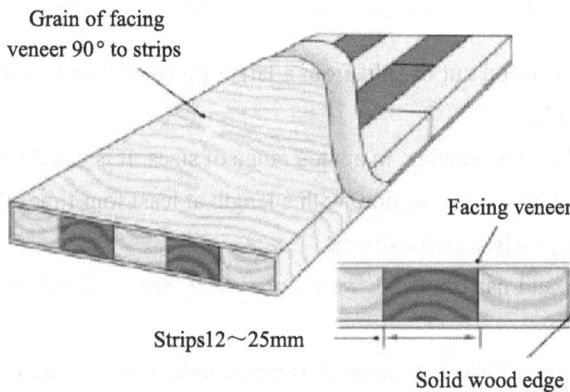

Grain of facing
veneer 90° to strips

Facing veneer

Strips12~25mm

Solid wood edge

Fig.7.27   Blockboard

(2) Particleboard

Particleboard is a panel product made by compressing small particles of wood while simultaneously bonding them with an adhesive (Fig.7.28). The many types of particleboards differ greatly with regards to the size and geometry of particles, the amount of adhesive used, and the density to which panels are pressed. The properties and potential uses of any board depend on these factors.

The major types of particles used for particleboards are:

① Shaving a small wood particle of indefinite dimensions produced when planning or jointing wood. It is variable in thickness and often curled.

② Flake a small particle of predetermined dimension produced by specialized equipment. It is uniform in thickness, with fiber orientation parallel to the faces.

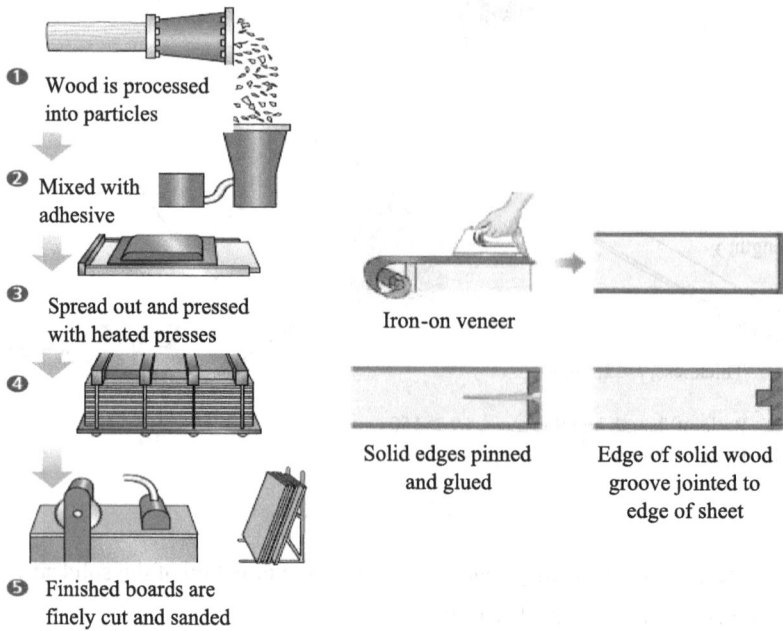

Fig.7.28　Making particleboard and edge covering

③ Chip a piece of wood cut or split from a larger piece of wood by a knife or hammer, as in a hammermill.

④ Sawdust produced by sawing, in a wide range of sizes. It is usually further refined.

⑤ Sliver nearly square cross-section, with a length at least four times the thickness.

⑥ Excelsior long, curly, slender slivers.

⑦ Strand a long and narrow flake, wafer, or veneer strip, 0.4~0.8mm thick, with parallel surfaces.

⑧ Wafer similar to a flake but larger. It is wider than a strand and nearly square. Usually over 0.5mm thick and over 25mm long. It may have tapered ends

Particleboard is often laminated with plastic or decorative wood veneer (such as oak, mahogany and ash) for furniture construction, worktops, bedrooms or kitchen units.

(3) Medium density fiberboard (MDF)

MDF is a panel product made primarily from wood fiber and bonded with synthetic resins to a density of 500~800kg/m^3 (Fig.7.29 and Fig.7.30). After particleboard, it is the most prominent nonstructural composite. Although it can be formed using a wet or a dry process, most of the production is formed dry. Like dry-processed hardboard, MDF is made from wood that has been reduced to individual fibers and fiber bundles with a binder added to affirm board strength. One of the keys to quality production of this product is the use of pressurized refiners, which produce pulp of a very low bulk density.

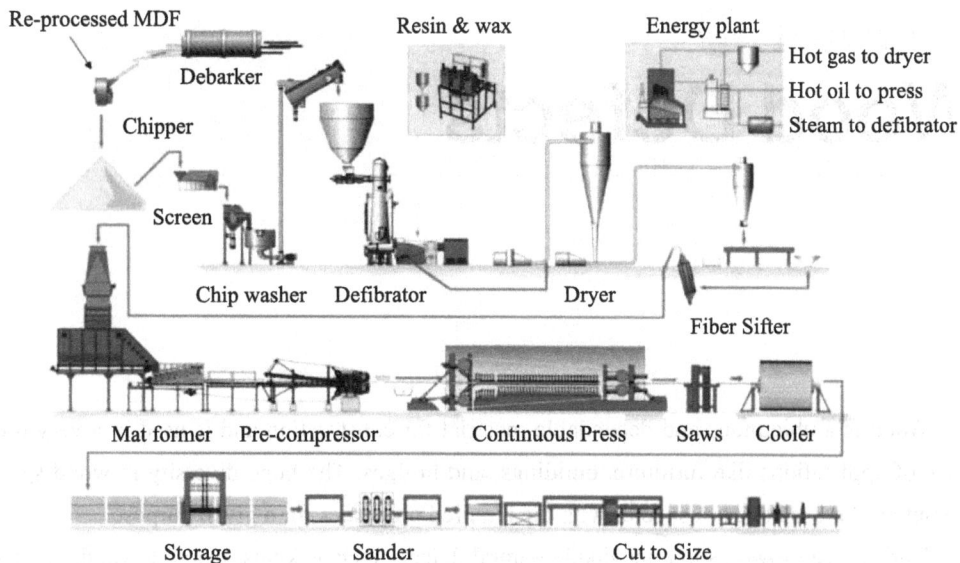

Fig.7.29 Making MDF

The first steps in making MDF are similar to those employed in manufacturing hardboard. Logs and other raw materials such as plywood and furniture trim or sawmill cut-off blocks are initially reduced to chips. The chips are then refined using thermomechanical pulping. In many cases, wax is added during pulping. Thereafter, the process closely resembles particleboard manufacturing. Special blending and forming machines are necessary for MDF because of the bulky low-density fiber. Most MDF processes use blow-line blending. There, the resin is injected into a pressurized pipe that contains the fiber from the refiner. UF resin is typically used as the binder with resin solids on the order of 8%~10% by weight. For some specialty products, the resin solids content can be 15% or greater. Then, the furnish is dried, which reduces subsequent press time and potential for blow-type delamination. Dried furnish is then formed dry and pressed into panels.

Fig.7.30 MDF product

Exercises

1. Describe the basic procedures for lumber manufacturing.

2. What is the advantage of wood seasoning?

3. What are engineered wood products (EWP)?

4. Describe the OSB production process.

5. What are the key differences between LVL and plywood?

6. What are the main production processes of GLT and CLT?

Chapter **❽** »»»
Wood Defects

Wood is a ubiquitous and dependable material for construction and is used in a very broad range of applications like furniture, buildings, and bridges. The huge diversity in wood species guarantees there is always a species with the required properties for a given purpose.

During tree growth, some inevitable natural defects such as knots, reaction wood, and taper occur. And artificial defects occur as a result of stress caused by poor stacking or seasoning. Additionally, given these originations from natural biomaterials, wood-based products are easily attacked by fungi and harmful insects, resulting in biodegradation. This section will discuss the common wood defects and the methods used to treat wood to avoid decreases in the value.

8.1　Wood Natural Defects

8.1.1　Knots

The seasonal addition of new wood results in a progressive layering over previously produced wood. As new growth increases the diameter of the main trunk, the branch bases become more and more deeply embedded in the trunk.

An examination of Fig.8.1 shows the living branch extending to the pith, the point at which most branches originate. The base of the branch is cone shaped, appearing as a tapered wedge when sectioned. The cone-shaped appearance arises from the fact that the cambium, which sheaths branches and the main trunk and moves farther from the embedded branch base. At the same time, the main trunk grows larger, preventing further diameter increases at this location. Because the main trunk and branch growth are simultaneous, an incorporation of living branches into

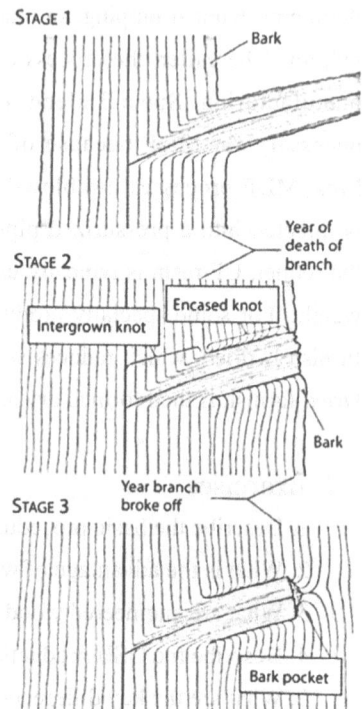

Fig.8.1　Development of a knot

the main trunk results in knots that become an integral part of the surrounding wood. Such knots do not become loose or fall out upon drying and are called *intergrown* or *tight knots* or *live knots*.

When a branch dies, its cambial layer also dies, stopping diameter growth throughout its length. The cambial layer of the main trunk or bole continues to grow, however, slowly encasing the dead branch in the process. Knots formed in this way are not an integral part of the surrounding wood and, if included in lumber, may fall out as drying takes place. These are called *loose* or *encased knots* or *dead knots*.

8.1.2 Reaction Wood

A reaction is a response to a triggering event. Reaction wood was, as such, appropriately named. This special kind of wood may be formed if the main trunk of a tree is tipped from the vertical, and it is known to regulate the orientation or angle of branches relative to the main trunk. It can also arise in response to accelerated growth and is reported to play a role in directing growing trunks toward openings in a forest canopy (Fig.8.2).

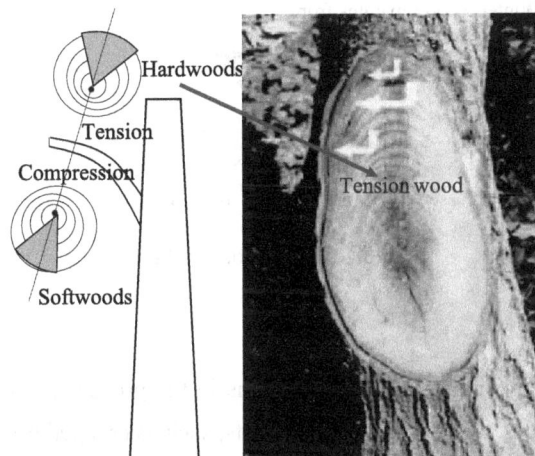

Fig.8.2   Reaction wood in hardwoods and softwoods

Reaction wood formed in hardwoods differs from that formed in softwoods. In softwoods it is termed *compression wood*; in hardwoods, it is called *tension wood*. However, the function of both reaction wood is the same, which is to bring the trunk or branch back to the original position.

8.1.3 Shakes

Shakes are splits in the end grain of the wood (Fig.8.3). They happen along the ray lines (radial shakes) and also along the annual rings (tangential shakes). These splits are caused by tension forces that built up in the wood while the tree is growing, or because of a rapid drying of the felled tree log before conversion. When a tree is felled, or before the log is converted, the forces within the log may be released. The weaker points in the wood break, and the split occurs.

Different types of splits (shakes) have different characteristics.

Radial shakes occur in the direction of the rays. These include heart shakes, star shakes and frost shakes. Wood splits inwards, as a result of very harsh weather conditions.

Tangential shakes occur in the direction of the annual rings. These occur in old age trees, in seasoning, or in strong wind. They include cup shakes where the split runs along the annual ring. Whereas a ring shake is when the split runs around the annual ring.

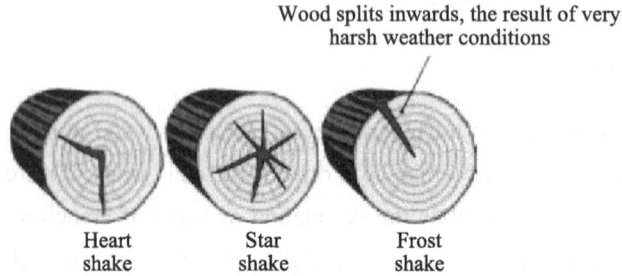

Wood splits inwards, the result of very harsh weather conditions

Heart shake　　Star shake　　Frost shake

Fig.8.3　Radial shakes

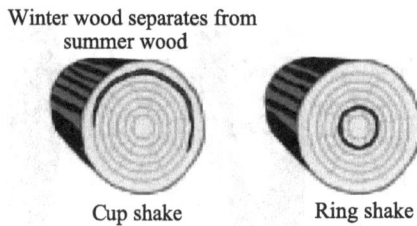

Winter wood separates from summer wood

Cup shake　　Ring shake

Fig.8.4　Tangential shakes

8.1.4　Trunk Shape Defect

Trees are affected by environmental conditions in the process of growth, therefore trunks may form abnormal shapes, called trunk shape defects, such as crooked trunk, stem taper, and burr (Fig.8.5).

Fig.8.5　Trunk shape defect

Crooked trunk: the axis (longitudinal center line) of the trunk is not in a straight line.

Stem taper: large diameter difference between the big end and the small end of the trunk or log.

Burr: one tumor formed by an abnormal growth of local tissue on the trunk.

8.1.5   Injury

Injuries refer to the scars caused by machinery, fire, bird and animal damage (Fig.8.6), mainly including mechanical damage, char, bird and animal injury, bark pocket, scar, knob, wind-breakage and tapping cut, etc.

Fig.8.6   Scar

8.2   Wood Biohazard Defects

8.2.1   Wood Biodegradation

By definition, sustainability involves making things last longer or extending their benefits in perpetuity. One of the primary ways to enhance forest sustainability is to extend the service life of wood products. If a product or structure's service life can be doubled, then in general, only half of the trees, or timberland area, are needed to provide that material. Wood protection and preservation enhances forest and environmental sustainability in this manner. In many cases, service lives are extended 5-to 10-fold. As such, the existing timberland base can provide a continually increasing amount of goods and services for a growing population.

Key to the satisfactory employment of wood and wood-based products as building materials is an understanding of the agents and conditions that can lead to decay or other forms of deterioration. Wood structures, when properly designed and constructed, can serve satisfactorily for hundreds of years. As with many naturally formed organic materials, however, wood may be subject to decay, fungal stains, insect infestation, fire, and surface weathering, all of which can greatly reduce the useful life of buildings and products.

There is no reason for wood deterioration to occur within a building where exposure to water can be controlled. Generally, wood that is used out of doors is subject to rain, or has contact with the ground, or to seawater will eventually decay or be attacked by insects or marine borers. The service life of wood products and structures is best enhanced when two factors are considered. First, building design and construction should be of a nature that minimizes direct wood and water contact, seeking to shed or drain water while avoiding water traps. Secondly, service life can be greatly extended by proper treatment or species selection. To avoid deterioration in buildings or to extend the life of wood materials used under severe exposure, users of wood products and structures must understand the conditions under which deterioration develops and take appropriate preventive measures.

Biological agents are the major causes of wood deterioration. Deterioration can result from a variety of organisms: fungi that cause staining, softening, and decay; marine borers, mainly small mollusks and crustaceans; insects, including termites, carpenter ants, carpenter bees, and a variety of wood-boring beetles; and bacteria that cause deterioration in water-stored logs and foundation piles. The greatest financial losses from biodeterioration result from decay fungi. These agents of biological deterioration are omnipresent. They are most active in tropical climates but also develop in temperate and colder regions.

Fungi that degrade wood are classified as decay, soft-rot, stain, or mold fungi according to the type of degradation they cause. *Decay fungi* cause significant softening or weakening of wood, often to the point that its physical and mechanical characteristics are completely destroyed. Wood so affected is referred to as rotten or decayed. *Soft-rot fungi* also weaken wood and most often attack wood that is very wet and usually penetrate rather slowly. Soft rots gradually degrade wood from the surface inwards. *Staining fungi* often create a bluish or blackish color when they inhabit wood and as such are detrimental to its appearance and value; yet they do not have a serious effect on the strength or the physical integrity of the wood. *Molds and mildews* occur only on exposed surfaces and may discolor products in use, such as house siding, but do not affect the strength. As early colonizers, molds are often an indicator that conditions are favorable to decay. The former two fungi decompose the cell walls of wood for their own survival and reproduction, causing wood decay and destruction. The latter two mainly absorb the substances in the cell cavities as nutrients, so they do not damage the cell walls.

The life cycle of fungi is composed of a growth period and a fruiting period. The damage done to wood occurs during the growth period. Under suitable conditions, the end of the mycelium of decay fungi can decompose enzymes and dissolve the wood cell walls into small pores, so the hyphae expand through these pores. However, the hyphae of thermochromic fungi enter into the cell lumens mainly through the pits on the cell walls, so it has little damage to wood.

Like many plants, fungi need four factors to grow: food, water, a favorable temperature, and sufficient oxygen.

8.2.2　Wood Decay

Decay fungi may be further classified as brown rots or white rots. The brown rots preferentially attack and consume cellulose and hemicellulose. Brown rots are most commonly noted in softwoods. Wood seriously degraded by these fungi will have an abnormally brownish or reddish color owing to the high concentration of residual modified lignin. Brown-rotted wood develops checks perpendicular to the grain when dried or handled break into cube-shaped pieces. Brown rots attack all layers of the cell wall but the cellulose-rich S_2 layer is often the first to be degraded. Brown rots cause dramatic strength loss in the early decay stages. Under a light microscope, early decay colonization appears as intermittent hyphal strands in and around the cell walls. Advanced decay often appears as missing and broken cell walls throughout. If the decay is well advanced, it may be difficult or impossible to slice a wood section to produce a microscope slide.

White-rot fungi have the ability to degrade both the lignin and cellulosic components of the cell, although the lignin is usually utilized at a somewhat faster rate. White rots may slightly change the color of wood, but more often they give wood a bleached or whitish color inherent to de-lignified cellulose. These fungi are most commonly noted in hardwoods. They typically erode the cell from the lumen outwards by decomposing successive layers of the cell wall. Thus, the cell wall becomes progressively thinner. However, the wood does not tend to shrink, check, or collapse as is often the case with brown rots. White-rotted wood usually retains its shape but may eventually become a fibrous spongy mass.

Soft-rot fungi also belongs to the Ascomycetes and Deuteromycetes classes. These organisms cause progressive degradation from the surfaces of wood inwards. The effects are slower to develop and less apparent than decay or staining fungi. However, in some wet service applications such as wood slats in cooling towers, transmission and distribution poles, and foundation and freshwater piling, structural failure by soft rots has been found. In the early stages, soft rot differs from other decay in that the affected surface(s) can be removed by scraping. The decomposition of cell walls by some types of soft rot fungi is characterized by long cavities in the longitudinal direction entirely within the secondary wall.

8.2.3　Wood Discoloration

Wood can be stained by chemicals (e.g., black bog oak), weathering, and fungi. Nearly all wood react differently to the acids or alkalis (soaps, detergents, etc.) with which it come in contact.

Fungal staining is caused by fungi infecting the wood and feeding on carbohydrates in the wood. These methods stain the wood blue, grey, black, pink, and white. Sometimes they occur on the surface and can be sanded off, but often they grow deep into the wood. Often the strength of wood is not affected, but its exterior appearance is poor. These stains occur due to poor seasoning. If moisture content is adjusted to 20% or below, the fungi cannot survive.

8.2.4 Insects

Wood products in use throughout the world are subject to infestation by a wide variety of termites (Fig.8.6) and beetles and by a few species of ants and bees. Some insects, particularly termites and wood-destroying beetles, cause great financial loss and are of concern in North America. A great number of other insects are important in specific regions of the world because of their damage to standing timber, logs, and wood products in use. Termites (biological order Isoptera) have the greatest economic impact by far.

Fig.8.6　Termites

8.3　Wood Artificial Defects

Wood surface damages during wood processing is called artificial defect, including sawing defects and drying defects. During wood sawing, there are mainly two kinds of sawing defects, namely, wane and defects, of sawn surface. Wood defects in seasoning or after seasoning are called drying defects.

Wane refers to the part of the log surface remained in the sawn timber, which is divided into waning arris and waning edge (Fig.8.7). A certain edge of the sawn timber that has not been cut in the width and thickness direction is the waning arris, and that in the length direction is waning edge.

Saw blade deviation during wood sawing result in defects of the sawn surface, including deep saw marks, snaking, rough saw cut, and saw kerf (Fig.8.8).

Artificial defects resulting from poor stacking or seasoning include checking and warping. Checks can be further divided into end splitting and honey checks (Fig.8.9). Meanwhile, warping includes twisting, cupping, bowing, and spring (Fig.8.10).

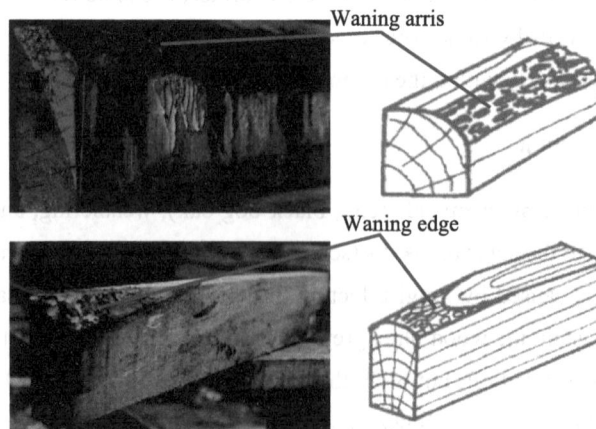

Waning arris

Waning edge

Fig.8.7　Wane

Deep saw marks

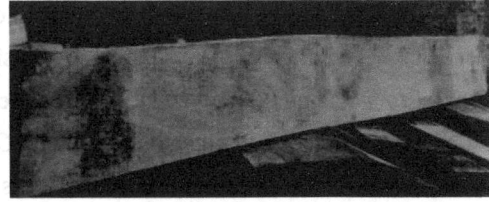
Snaking

Rough saw cut

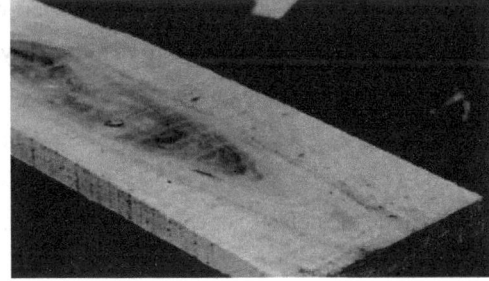
Saw kerf

Fig.8.8 Defects of sawn surface

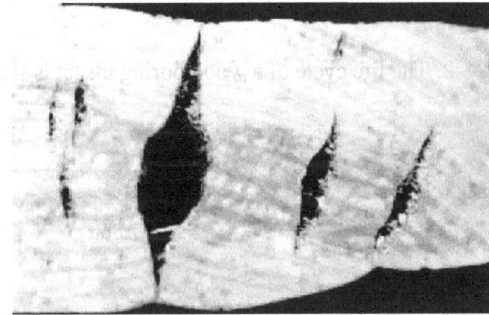

Fig.8.9 Checks

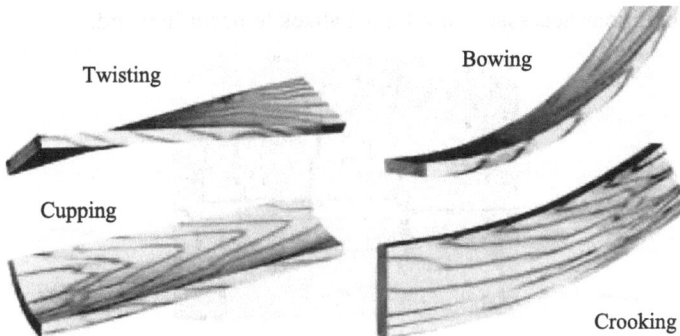
Twisting

Bowing

Cupping

Crooking

Fig.8.10 Warping

End-splitting occurs if the ends of boards dry out too quickly, due to exposure to the sun or heat. It is very common in naturally seasoned timber. Oftentimes, painting or treating the ends during seasoning can prevent it.

Where the inside of the wood splits is called honeycomb checks. This happens when wood dries quickly and the inside dries before the outside, tearing the wood fibers.

Twisting occurs when two edges of wood remain straight, but the faces are distorted as if the two ends were twisting in opposite directions. *Cupping* is a type of shrinkage that forms a curve if you view a plank from either end. A *bow* takes the form of a bend along the length of the board. Both cupping and bowing defects are often due to the incorrect stacking of the boards during seasoning, where the stickers are too far apart or perhaps not directly above each other. It may also be as a result of poor stacking in a timber yard. *Crooking* refers to the lateral bending of the timber edge along the longitudinal direction.

▶ Exercises

1. Name the defects shown in the diagrams.

2. The life cycle of a wood boring insect is shown. At which stage does the most damage occur?

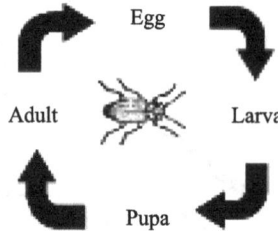

3. List the conditions necessary for a fungal attack to occur in wood.

4. Planks that are seasoned too quickly can develop several types of defect. Using notes and sketches describe any two kinds of these defects (JC, OL 2006).

5. Describe a wane.

6. Recommend a suitable preservative for a garden shed and describe a method that could be used to apply it.

参考文献 REFERENCES

曹金珍，2006. 国外木材防腐技术和研究现状 [J]. 林业科学，42(7): 120–126.

柴宇博，2015. 木材乙酰化以及作用机制研究 [D]. 北京：中国林业科学研究院.

陈幸良，巨茜，林昆仑，2014. 中国人工林发展现状、问题与对策 [J]. 世界林业研究，27(6): 54–59.

程汉亭，刘书伟，黎明，2014. 名贵硬木树种及木材识别 [M]. 北京：中国农业科学技术出版社.

崔会旺，杜官本，2008. 木材阻燃研究进展 [J]. 世界林业研究，21 (3): 43–48.

丁涛，2019. 木材科学与工程专业英语 [M]. 北京：中国林业出版社.

丁涛，蔡家斌，耿君，2015. 欧洲木材热处理技术的研究及应用 [J]. 木材工业，29(5): 29–39.

樊承谋，王永维，潘景龙，2009. 木结构 [M]. 北京：高等教育出版社.

符韵林，李英健，2019. 红木鉴 [M]. 北京：中国轻工业出版社.

顾炼百，丁涛，江宁，2019. 木材热处理研究及产业化进展 [J]. 林业工程学报，4(4): 1–11.

郭晓磊，林雨斌，滕雨，等，2016. 正交胶合木生产的关键设备 [J]. 木材加工机械，27(2): 55–58.

国家林业和草原局，2019. 中国森林资源报告 (2014—2018)[M]. 北京：中国林业出版社.

国家林业局森林资源管理司，2013. 全国森林资源统计：第八次全国森林资源清查[M]. 北京：中国林业出版社.

何敏娟，Lam F，杨军，等，2008. 木结构设计 [M]. 北京：中国建筑工业出版社.

贾福根，宋高嵩，2016. 土木工程材料 [M]. 北京：清华大学出版社.

江进学，李建章，梅超群，2009. 木质材料阻燃改性研究进展 [J]. 应用化工，38 (8): 1203–1206.

蒋明衍，陈奶荣，林巧佳，2013. 木材防腐的研究进展 [J]. 福建林业科技，40(1): 207–213.

黎云昆，2017. 木材及木文化刍议 [J]. 西南林业大学学报 (社会科学), 1(1): 33–38.

李坚，2013. 木材保护学 [M]. 北京：科学出版社.

李坚，2014. 木材科学 [M]. 北京：科学出版社.

李涛，蔡家斌，周定国，2013. 木材热处理技术的产业化现状 [J]. 木材加工机械 (5): 50–53.

李霞，余荣卓，2018. 森林文化 [M]. 北京：中国林业出版社.

李晓琴，徐煜，2003. 软腐——一种特殊的木材腐朽 [J]. 林业科技，28(3): 46.

李征，罗晶，何敏娟，2018. 我国装配式木结构标准体系现状及完善建议 [J]. 工程建设标准化 (10): 67–72.

联合国粮食及农业组织, 2020. 2020 年世界森林状况：森林、生物多样性与人类 [M]. 罗马：联合国粮食及农业组织.

刘鸿文, 2005. 简明材料力学 [M]. 北京：高等教育出版社.

刘伟庆, 杨会峰, 2019. 现代木结构研究进展 [J]. 建筑结构学报 (2): 16–43.

罗建举, 2015. 木与人类文明 [M]. 北京：科学出版社.

骆介禹, 骆希明, 2003. 纤维素基质材料阻燃技术 [M]. 北京：化学工业出版社.

彭亮, 许柏鸣, 2016. 家具设计与工艺 [M]. 北京：高等教育出版社.

阙泽利, 李哲瑞, 王菲彬, 等, 2017. 中高层木结构用正交胶合木 (CLT) 在欧洲的研究与发展现状 [J]. 建筑结构, 47(2): 75–80.

孙芳利, Prosper N, 吴华平, 等, 2017. 木竹材防腐技术研究概述 [J]. 林业工程学报, 2(5): 1–8.

王冬米, 2016. 走进森林 [M]. 北京：中国农业科学技术出版社.

王飞, 刘君良, 吕文华, 2017. 木材功能化阻燃剂研究进展 [J]. 世界林业研究, 30 (2): 62–66.

王灵燕, 聂玉静, 陈争骥, 等, 2016. 我国木材阻燃研究现状及发展趋势 [J]. 浙江林业科技, 36 (5): 82–88.

王清文, 李坚, 1999. 木材阻燃工艺学原理 [M]. 哈尔滨：东北林业大学出版社.

王世襄, 2013. 明式家具研究 [M]. 上海：三联书店.

王艳伟, 孙伟圣, 杨植辉, 等, 2014. 木材高温热处理技术的研究进展及展望 [J]. 林业机械与木工设备, 42(9): 8–11.

文博, 2013. 实木在现代建筑室内装修中的应用研究 [D]. 南京：南京工业大学.

吴楚材, 吴章文, 胡卫华, 等, 2013. 森林——人类健康的摇篮 [M]. 北京：中国旅游出版社.

熊海贝, 宋依洁, 戴颂华, 等, 2018. 装配式 CLT 建筑从模型到建造 [J]. 建筑结构, 48(10): 7–12.

徐有明, 2006. 木材学 [M]. 北京：中国林业出版社.

殷亚方, 2016. 常见贸易濒危与珍贵木材识别手册 [M]. 北京：科学出版社.

张婷婷, 孙巧, 孙雪敏, 等, 2017. 正交胶合木的研究现状及国产化展望 [J]. 林业机械与木工设备, 45(1): 4–7.

赵广杰, 2021. 木文明基本架构 [J]. 林产工业, 58(4): 76–80.

中国资源科学百科全书编辑委员会, 2000. 中国资源科学百科全书 [M]. 东营：石油大学出版社.

周京南, 2016. 木性药考：中国传统家具用材的药用价值研究 [M]. 北京：中国林业出版社.

Buck D, Wang X A, Hagman O, et al., 2016. Bending properties of cross laminated timber (CLT) with a 45 alternating layer configuration[J]. BioResources, 11(2): 4633–4644.

Chen Y, 2011. Structural performance of box based cross laminated timber system used in floor applications[D]. Vancouver: University of British Columbia.

Chen Y, Lam F, 2012. Bending performance of box–based cross–laminated timber systems[J]. Journal of Structural Engineering, 139(12): 4013006.

Davim J P, Aguilera A, Henriques A, 2017.Wood welding without adhesives: Materials,

Manufacturing and Engineering [M]. Berlin: De Gruyter Press.

Fast P, Gafner B, Jackson R, et al., 2016. Case study: an 18 storey tall mass timber hybrid student residence at the University of British Columbia, Vancouver[C]. WCTE 2016—World Conference on Timber Engineering, Vienna, Austria.

Filley T, Blanchette R, Simpson E, et al., 2001. Nitrogen cycling by wood decomposing soft—rot fungi in the "King Midas tomb," Gordion, Turkey [J]. PNAS, 98 (23): 13346–13350.

Karacabeyll E, Gagnon S, 2019.CLT Handbook: cross—laminated timber[M].Vancouver: FPInnovations.

Michael C, 2010. Wood Materials Technology [M]. 4th ed. Dublin: The Educational Company of Ireland.

Richardson B, 1993. Wood Preservation[M]. London: E&FN SPON.

Schubert S, Gsell D, Steiger R, et al., 2010. Influence of asphalt pavement on damping ratio and resonance frequencies of timber bridges[J]. Engineering Structures, 32(10): 3122–3129.

Shmulsky R, Jones P, 2019. Forest Products and Wood Science：An Introduction [M]. Hoboken: John Wiley & Sons Ltd.

Strobel K, 2016. (Mass) Timber: Structurally Optimized Timber Buildings[D]. Seattle: University of Washington.

Wei P, Wang B, Li H, et al., 2019. A comparative study of compression behaviors of cross—laminated timber and glued—laminated timber columns[J]. Construction and Building Materials (222): 86–95.

Zimmermann M, 1983. Xylem Structure and the Ascent of Sap [M]. New York: Springer.